Lecture Notes in Economics and Mathematical Systems 512

Founding Editors:

M. Beckmann
H. P. Künzi

Managing Editors:

Prof. Dr. G. Fandel
Fachbereich Wirtschaftswissenschaften
Fernuniversität Hagen
Feithstr. 140/AVZ II, 58084 Hagen, Germany

Prof. Dr. W. Trockel
Institut für Mathematische Wirtschaftsforschung (IMW)
Universität Bielefeld
Universitätsstr. 25, 33615 Bielefeld, Germany

Co-Editors:

C. D. Aliprantis, Dan Kovenock

Editorial Board:

P. Bardsley, A. Basile, M.R. Baye, T. Cason, R. Deneckere, A. Drexl,
G. Feichtinger, M. Florenzano, W. Güth, K. Inderfurth, M. Kaneko, P. Korhonen,
W. Kürsten, M. Li Calzi, P. K. Monteiro, Ch. Noussair, G. Philips, U. Schittko,
P. Schönfeld, R. Selten, G. Sorger, R. Steuer, F. Vega-Redondo, A. P. Villamil,
M. Wooders

Springer
*Berlin
Heidelberg
New York
Barcelona
Hong Kong
London
Milan
Paris
Tokyo*

Wei-Bin Zhang

An Economic Theory of Cities

Spatial Models with Capital, Knowledge, and Structures

 Springer

Author

Prof. Wei-Bin Zhang
Ritsumeikan Asia Pacific University
Jumonjibaru, Beppu-Shi
874-8577 Oita-ken, Japan

Cataloging-in-Publication data applied for

Die Deutsche Bibliothek - CIP-Einheitsaufnahme

Zhang, Wei-Bin:
An economic theory of cities : spatial models with capital, knowledge, and structures / Wei-Bin Zhang. - Berlin ; Heidelberg ; New York ; Barcelona ; Hong Kong ; London ; Milan ; Paris ; Singapore ; Tokyo : Springer, 2002
 (Lecture notes in economics and mathematical systems ; 512)
 ISBN 3-540-42767-8

ISSN 0075-8450
ISBN 3-540-42767-8 Springer-Verlag Berlin Heidelberg New York

This work is subject to copyright. All rights are reserved, whether the whole or part of the material is concerned, specifically the rights of translation, reprinting, re-use of illustrations, recitation, broadcasting, reproduction on microfilms or in any other way, and storage in data banks. Duplication of this publication or parts thereof is permitted only under the provisions of the German Copyright Law of September 9, 1965, in its current version, and permission for use must always be obtained from Springer-Verlag. Violations are liable for prosecution under the German Copyright Law.

Springer-Verlag Berlin Heidelberg New York
a member of BertelsmannSpringer Science+Business Media GmbH

http://www.springer.de

© Springer-Verlag Berlin Heidelberg 2002
Printed in Germany

The use of general descriptive names, registered names, trademarks, etc. in this publication does not imply, even in the absence of a specific statement, that such names are exempt from the relevant protective laws and regulations and therefore free for general use.

Typesetting: Camera ready by author
Cover design: *design & production*, Heidelberg

Printed on acid-free paper   SPIN: 10853976   55/3142/du   5 4 3 2 1 0

Acknowledgements

I completed this book at the Ritsumeikan Asia Pacific University (APU) in Japan. Helpful in many ways at the early stage of my research and settlement at the APU were Professors Yuji Jido, Kenichi Nakagami, and Issei Nakanishi. I am grateful to the anonymous referees for valuable comments. I would like to thank Economics Editor Dr. Werner A. Müller and Economics Editorial Christiane Beisel for effective co-operation.

Chapters 2 to 11 are based on my published or unpublished manuscripts. Grateful acknowledgment is made to the following sources for the use of my published materials:

Chapter 2 for Zhang, W.B. (1996d);
Chapter 3 for Zhang, W.B. (1993a, 1993b);
Chapter 5 for Zhang, W.B. (1994e);
Chapter 6 for Zhang, W.B. (1993c);
Chapter 7 for Zhang, W.B. (1996c, 1998c);
Chapter 8 for Zhang, W.B. (1997a);
Chapter 9 for Zhang, W.B. (1993d);
Chapter 10 for Zhang, W.B. (1996b, 1998a,1998b);
Chapter 11 for Zhang, W.B. (1994a,1994b).

This book is based on a part of my research supported by APU Academic Research Subsidy FY2000 over the last year, which is gratefully acknowledged.

Preface

Over more than two centuries the development of economic theory has created a wide array of different concepts, theories, and insights. My recent books, *Capital and Knowledge* (Zhang, 1999) and *A Theory of International Trade* (Zhang, 2000) show how separate economic theories such as the Marxian economics, the Keynesian economics, the general equilibrium theory, the neoclassical growth theory, and the neoclassical trade theory can be examined within a single theoretical framework. This book is to further expand the framework proposed in the previous studies.

This book is a part of my economic theory with endogenous population, capital, knowledge, preferences, sexual division of labor and consumption, institutions, economic structures and exchange values over time and space (Zhang, 1996a). As an extension of the *Capital and Knowledge*, which is focused on the dynamics of national economies, this book is to construct a theory of urban economies. We are concerned with dynamic relations between division of labor, division of consumption and determination of prices structure over space. We examine dynamic interdependence between capital accumulation, knowledge creation and utilization, economic growth, price structures and urban pattern formation under free competition. The theory is constructed on the basis of a few concepts within a compact framework. The comparative advantage of our theory is that in providing rich insights into complex of spatial economies it uses only a few concepts and simplified functional forms and accepts a few assumptions about behavior of consumers, producers, and institutional structures.

This book constructs a theoretical framework that would permit valid generalizations from one special modeling structure to another, and would deepen our understanding of economic evolution. It is a part of my broad approach to revealing complex of economic evolution (Zhang, 1991d, 1996a, 1999, 2000). I wish that the reader would appreciate this book within the grand framework that I have made great efforts to construct.

Wei-Bin Zhang
Beppu-Shi, September 2001

Contents

1 Introduction 1
  1.1 Von Thünen's Theory 1
  1.2 Classical Location Theory 3
  1.3 The Alonso Model and its Extensions 5
  1.4 Imperfect Competition 11
  1.5 Spatial Agglomeration 12
  1.6 Spatial Structures with Population and Knowledge 13
  1.7 City Systems 14
  1.8 Nonlinear Spatial Economic Dynamics 16
  1.9 The Purpose and Structure of the Book 17

2 Urban Growth with Housing and Spatial Structure 21
  2.1 Urban Growth with Housing Production 21
  2.2 The Dynamics in the Terms of $K(t)$ 29
  2.3 Equilibrium and Stability 31
  2.4 The Impact of the Population on Economic Geography 35
  2.5 The Propensity to Hold Wealth and the Equilibrium
       Structure 37
  2.6 On Extensions of the Basic Model 40
  Appendix 40
  A.2.1 Proving Proposition 2.1 40

3 Spatial Pattern Formation with Capital and Knowledge ... 42
  3.1 The Urban Dynamics 43
  3.2 Equilibrium and Stability 48
  3.3 The Knowledge Accumulation Parameters 50
  3.4 The Impact of Government Intervention in Research 52
  3.5 The Working Conditions of Scientists 55
  3.6 On Knowledge Creation and Spatial Economic Evolution 55
  Appendix 56
  A.3.1 Expressing the Variables in Terms of K and Z 56
  A.3.2 The Proof of Proposition 3.2.1 59

4 Urban Structure with Growth and Sexual Division of Labor 61
  4.1 Growth with Sexual Division of Labor and Location 62

4.2	The Spatial Equilibrium Structure	67
4.3	Sexual Productivity Differences and Economic Structure	69
4.4	Remarks	72
Appendix		72
A.4.1	Proving Proposition 4.2.1	72

5 Dynamic Urban Pattern Formation with Heterogeneous Population ... 76
- 5.1 The Urban Growth with Two Groups ... 77
- 5.2 Separation of the Groups' Residential Location ... 83
- 5.3 Economic Equilibrium and Stability ... 86
- 5.4 The Impact of Savings Rates ... 89
- 5.5 The Impact of Human Capital ... 92
- 5.6 On Urban Evolution with Multiple Groups ... 93
- Appendix ... 94
- A.5.1 Proving Lemma 5.3.1 ... 94
- A.5.2 Proving Lemma 5.3.2 ... 97

6 Two-Group Spatial Structures with Capital and Knowledge 100
- 6.1 The Spatial Economy with Heterogeneous Population ... 100
- 6.2 Temporary Urban Pattern ... 105
- 6.3 Long-Run Equilibria and Stability Conditions ... 108
- 6.4 The Impact of Creativity ... 109
- 6.5 The Impact of the Population ... 113
- 6.6 On Urban Evolution with Heterogeneous Population ... 114
- Appendix ... 114
- A.6.1 Proving Lemma 6.2.2 ... 114
- A.6.2 Proving Proposition 6.3.1 ... 116

7 Urban Growth and Pattern Formation with Preference Change ... 118
- 7.1 The Model ... 118
- 7.2 Properties of the Dynamic System ... 126
- 7.3 Human Capital and the Economic Structure ... 129
- 7.4 On Preference Change and Urban Pattern ... 132
- Appendix ... 133
- A.7.1 Proving Lemma 7.2.1 ... 133

8 Urban-Rural Division of Labor with Spatial Amenities 135
- 8.1 The Spatial Structure with Urban-Rural Areas ... 135
- 8.2 Spatial Equilibria ... 140
- 8.3 Amenities and Economic Geography ... 141
- 8.4 On Spatial Equilibrium Structure ... 144
- Appendix ... 145
- A.8.1 Proving Proposition 8.2.1 ... 145

9 Spatial Equilibrium with Multiple Cities ... 150
9.1 The Model ... 151
9.2 Urban Equilibria ... 156
9.3 The Impact of Amenities upon the Urban Structure ... 159
9.4 Concluding Remarks ... 161
Appendix ... 162
A.9.1 Proving Proposition 9.2.1 ... 162

10 Growth with International Trade and Urban Pattern Formation ... 166
10.1 The Model ... 167
10.2 The Dynamic Properties of the Trade System ... 172
10.3 Country 1's Propensity to Own Wealth ... 177
10.4 Country 1's Working Efficiency ... 181
10.5 On International Trade and Spatial Structures ... 181
Appendix ... 182
A.10.1 Proving Proposition 10.2.1 ... 182
A.10.2 Proving Proposition 10.2.2 ... 183

11 Nonlinear Dynamics of a Multi-City System ... 185
11.1 An Isolated Island Economy ... 186
11.2 Economic Geographic Cycles in the Isolated Island Economy ... 190
11.3 Aperiodic Oscillations in the J-Island Economy ... 195
11.4 On Spatial Chaos ... 198
Appendix ... 198
A.11.1 The Dynamics of Capital and the Population ... 198

12 Further Issues on Cities ... 201

Bibliography ... 204

Author Index ... 217

1 Introduction

When we look at a collection of city maps, we tend to be amazed by the enormous variety of shapes that social and economic mechanisms create under constraints of the nature. When we follow history of great cities, we feel strongly about dynamics of urban life. A single city shows different economic geographical patterns when it grows in different historical conditions. Cities grow, stagnate, and decline, depending on internal evolutionary mechanisms and their relations with the rest of the world. Economists have long been interested in searching for the causes and effects of urban growth. However, consistency and connectivity are weak among these approaches. It is reasonable to ask whether it is possible to build a general framework within which the varied urban issues and economic principles addressed in the traditional approaches can be examined in a consistent manner. The purpose of this book is to make an initial step towards constructing such a 'general' urban economic theory.

This book is concerned with economic dynamics, land use and spatial economic structures. Since land is immobile and cannot be augmented at a given location, it limits intensity of human activities and affects cost of human interactions. Irrespective of their importance in shaping economic activities in geographical space, aspects of dynamic spatial economies have received only scant attention in modern theoretical economics. One reason for this appears to be that it is difficult to take account of economic space within the well-established analytical frameworks. It has become clear that addition of explicit consideration of spatial factors such as neighborhood effects and transportation costs may cause some essential changes in basic results of general economic theory in which spatial effects are explicitly omitted. As pointed out by Ponsard (1983), space had either been ignored or treated superficially in the Anglo-Saxon tradition of economic theory. The major original spatial economists, such as von Thünen (1826), Weber (1909), Christaller (1933), and Lösch (1938), were all outside the Anglo-Saxon tradition. It is only in recent years that space has found its way into mainstreams of economics.

1.1 Von Thünen's Theory

Johann Heinrich von Thünen is the father of location theory and a discoverer of the marginal productivity theory of distribution. He was born on June 24, 1783, on his father's estate in the Grand Duchy of Oldenburg. He lived in Germany from 1783 until 1850. As pointed out by Blaug (1985), the history of location theory begins with the publication of von Thünen's *Der Isolierte Staat* (The Isolated State) in 1826. He was living in an agricultural economy. A main character of agricultural activities is that land is used extensively. von Thünen was concerned with the problem of agricultural activities which center on the competition of land use. He provided the classical analysis of allocation of land among competing agricultural activities. He started to construct the isolated state economy as follows (Von Thünen, 1826):

> Consider a very large town in the center of a fertile plain which does not contain any navigable rivers or canals. The soil of the plain is assumed to be of uniform fertility which allows cultivation everywhere. At a great distance the plain ends in an uncultivated wilderness, by which this state is absolutely cut off from the rest of the world.
> This plain is assumed to contain no other cities but the central town and in this all manufacturing products must be produced; the city depends entirely on the surrounding country for its supply of agricultural products.
> All mines and mineral deposits are assumed to be located right next to the central town.
> It was assumed that there was only one form of transportation, the horse and wagon, operated by the farmer at his own expense. There were no multiple freight rates depending on the commodities sipped.

The question now is: How under these circumstances will agriculture be developed and how will the distance from the city affect agricultural methods when these are chosen in the optimal manner? He showed that the town would be surrounded by agricultural rings: each ring cultivates a specific crop associated with the highest bid rent over the ring. The bid rent is the revenue minus the cost of labor and the cost of transportation. Differences in land use and agricultural production were the result of the types and quantities of agricultural products needed in the city, the technology employed in the production and transportation of such commodities, and the endeavor of each farmer to maximize his land rent by producing commodities. Various kinds of agricultural products are grown in concentric circles around the city, the exact location at which each product is raised being determined by the cost of transporting it to the city. For instance, in the area near to the city vegetables and fresh milk are produced and land cultivation is intensive because of the high price of land. Farther from the city, wheat is produced by the enclosure system. In the outermost circles the land is used for grazing.

Using the simple model, von Thünen examined many important issues in spatial analysis such as the effect of location on prices and land rent, the effect of urban demand fluctuations on the corresponding agricultural area, and interactions between the city and its hinterland. The model explains a number of important issues related to agricultural production locations in a simple manner. It explains the existence of agricultural specification even in the absence of climatic and fertility differences and the decrease of rent with distance from markets. It explains the relationship among distance of farms from the market, prices received by farmers for their products and land rent. The price which a farmer obtained for a given unit of commodity was equal to its price at the market minus the cost of shipping it to the market. The cost of transportation increased with distance from the market. The land rent was dependent on location relative to the market. The land rent nearest the city was the highest. The land rent decreased as location was further away from the city. It is a fertile model in the sense that it can be extended and modified in multiple ways. A key concept in land use theory is the concept of land rent. Ricardo and von Thünen wrote about agricultural land rent. Ricardo emphasized fertility differences while von Thünen was concerned with location differences. Modern urban economics was strongly influenced by von Thünen's work. As most of classical works, von Thünen's book is characterized by being insightful into many important issues.

1.2 Classical Location Theory

Classical location theory is often concerned with following two problems (Norman, 1979). The first one is that given the locations of all other economic agents, how should a particular agent such as a firm or industry be located to minimize the cost of serving a known, fixed demand? This problem is solved by the so-called least cost approaches to location theory. The second question is that given that firms are in direct competition with each other, how will they locate and what market areas will they control, given knowledge of demand conditions. This problem is analyzed by the central place theories and the interdependence theories.

The least cost approach was first developed by Weber (1909). He tried to find out what are the general factors of location for industry and what's role these factors play in affecting industrial location. He developed his location theory mainly basing on the following assumptions: (1) the firm's technology exhibits constant returns to scale; (2) production factors are available in unlimited supply at fixed prices independent of location; (3) these factors are either available everywhere in the market or located in a few fixed sources; (4) demand is known and fixed in terms of space and amount; and (5) transport costs for each commodity are directly proportional to weight and distance transported. The optimum location of a firm is defined as a search for the minimization of production and transportation costs. Under these assumptions, he showed how a firm would choose the location in order to minimize the cost of serving a particular consumption site. In the case of two inputs located at two different points and one product market located at the third point, Weber showed that the optimal location lies within a triangle formed by linking the product market point with two input points and two input points with each other. He also mentioned that the optimum location is found by taking into consideration the relative strength of two material pull forces and the product market pull force. Weber set up the paradigm for location of facilities (such as plants, warehouses, military bases, schools, waste material dumps, fire engine depots, hospitals, administrative buildings, and department stores) based on the minimization of transportation costs. It should be noted that Weber also tried to analyze locational issues when there are economies of agglomeration. He considered many subjects in location theory.

Weber's lasting influence on the subsequent developments of location theory did not begin immediately after he created his theory. It was only after 1960 that location theory emerged as a discipline. It is Isard (1956) who began to extend Weber's least cost theory. He incorporated the substitution principle into location theory. Moses

(1958) introduced neoclassical production function into location theory. Assuming that a firm chooses its plant location within the Weber-Moses triangle and the plant location is constrained to remain at a specified distance from the output market, he showed that the location decision of a firm is independent of the output level if the firm's production function is linearly homogeneous. Moreover, Moses noted that profit maximization requires a proper adjustment of output, input combination, location and price. He considered the optimum location to be finally dependent on base prices on inputs, transportation rates on inputs and on the final product, the geographical position of materials and markets, the product function, and demand function. Sakashita (1987) further examined the Moses problem by considering the location decision of a firm in a straight line, which connects two markets in space. He showed that the location decision of a cost-minimizing firm is independent of the output level if the production function is linearly homogenous. He also examined the impact of demand on the location decision of a profit-maximizing monopolistic firm. He proved that the optimum location of a firm is independent of the demand function if and only if the production function is linearly homogeneous. There are many attempts to extend Sakashita's analysis (e.g., Shieh and Mai, 1997). For instance, Shieh and Mai explicitly incorporated monopoly market structure into the Weber-Moses triangle model. They examined the effects of demand on the production and location decisions of a firm. By employing a two-stage approach, it is shown that the optimum location is not independent of the demand function if the production expansion path is linear through the origin and the distance of plant location from the product market is given. In the model the linearly homogenous production function is not sufficient to ensure that the optimum location is independent of the demand function if the distance from plant location is a choice variable. It is found that when demand is price-dependent, the optimum location for a firm should be located at the point of maximum profit rather than minimum cost (Beckmann, 1987). But most of theoretical studies on the location theory of a firm are to minimize the cost of producing and selling a given level of output. Little attention has been given to cases in which the objective of a firm is to maximize the profit. There are exceptions. For instance, In Shieh (1989), the firm's optimum location is determined in the context of the maximum profit rather than the minimum cost.

Christaller (1933) created a theory about market areas. He explained the laws that determine the number, size, function and spacing of settlements. His theory is about the location of tertiary activities, which stands alongside the work of von Thünen on the location, and the location theory of secondary activities by Weber. It is Lösch who first provided reasons why economic activities tend to agglomerate over an otherwise featureless plain. The central place theory by Lösch (1938, 1940) is based on the following assumptions: (1) a firm maximizes its profit in choosing location; (2) there are no abnormal profits in economic activities, which are open to everyone. Different from Weberian theory which is concerned with a heterogeneous market area with consumption fixed and concentrated at distinct points, Lösch was concerned with an unbounded, homogeneous plain over which consumers are evenly distributed. Rather than a fixed demand, Lösch assumed a downward-sloping demand curve for whatever is being produced. He built the theory by assuming a homogeneous plain containing regularly distributed self-sufficient farms. First, consider one farmer who wishes to produce beer and sell them to the consumers with an identical demand curve. The profit maximization provides the brewer a circular market area either with surplus profits or an extensive market area, which will attract other producers. Other producers compete away the demand until every producer gets the normal profit. It is through this kind of competition that the spatial shape of the individual market areas is determined. Lösch showed that a system of hexagonal markets would emerge. A modern proof of the spatial pattern formation was given by Bollobas and Stern (1972). In this system the size of each hexagon is determined by the relationship between production, demand and transport conditions. Since an economy consists of multiple industries and each industry forms its own system of hexagonal markets, different industries will give rise to different sizes of hexagons. Lösch showed how a hierarchy of production centers will emerge if nets of hexagons of different sizes are superposed in an ordered manner. By this way, Lösch was able to provide an explanation for urban formation and hierarchical systems of cities.

We should consider these two approaches as two extreme cases. Weber ignored the demand factor in his locational analysis; Lösch omitted spatial variation in cost conditions. Isard (1956, 1960) made attempts to integrate the two approaches some years ago.

1.3 The Alonso Model and its Extensions

Von Thünen developed a theory of the location of agricultural products in concentric rings around a central market. Isard (1956) noticed that the ideas of von Thünen could be reinterpreted in the context of urban land-use. Since then, there are some works, which mark a conceptual transition between the earlier work of land economists and modern urban economists (Beckmann, 1957, Mohring, 1961, Wingo, 1961, Muth, 1963). The availability of a well-developed microeconomic theory provided the theoretical foundations for the growth of modern urban economics. Beckmann (1957) studied the determination of equilibrium residential land rents and quantities in a monocentric city, where all employment and services are concentrated in the CBD (central business district) surrounded by a residential area. But it may be argued that the development of modern urban economics has been strongly influenced by the work of Alonso (1964). In his bid-rent theory, Alonso pioneered the adaptation of von Thünen's work on an urban context. The central market was replaced by a central business district, and agricultural products by alternative urban lands. According to Papageorgiou and Pines (1999), Alonso's greatest contribution was his proposal for the matching between spatial analysis and microeconomic theory that was necessary for the development of modern urban economics. Similar to the Solow-Swan model for neoclassical growth model, the Alonso model provides a simple mathematical structure based on which many articles have been published. The model has inspired further development of urban economics mainly because of its logical consistency, simplicity, possible extensions and rich implications for important phenomena. Since the publication of Alonso's seminal work, urban economic has become an established field through the works of urban economists.

The Alonso model is concerned with urban land use and market land prices. In its simple form, the model of monocentric city assumes that all economic activities are concentrated in central business districts, which are surrounded by residential suburbs. It involves a density of consumers, identical with respect to income and tastes. Preferences are defined over the consumption of a composite good, which is found at the CBD and land. Land is considered as a commodity. Each consumer must occupy land at one and only one location. Since the disposable income of consumers varies with distance from the center due to differences in transportation costs, so does consumption. In equilibrium consumption and prices are such that everyone has the same utility level. The model explains pattern formation of the residential land use around the CBD. The price of land, the density of land use, and the equilibrium locations of the urban population are endogenously determined by the model. The key role of analysis in this approach is the concept of 'bid rent', which represents the maximum rent each participant in the market can pay at each location. As a result of the different possible uses of a location, there is a whole set of bid rents for any given location. The equilibrium rent is the maximum of the bid rents, which is geographically represented as the upper envelope of bid rent curves.

We now formally illustrate the Alonso model. For simplicity, we accept a simplified version of the Alonso model provided by Fujita (1989). There are many factors affecting households' decision-making on residential location. Households should consider accessibility of residential locations to shopping centers, working place, friends and transport systems. Land price and space availability constraint possibility of locational choice. One may also have to include local amenities such as traffic noise and greenness in analyzing residential location. Faced with time and budget constraints, households have to weigh all these factors in making the decision. The model in this section takes account of two locational factors, accessibility and space.

We assume that the economy has a single household type. There are N identical households in the city. All job opportunities are located in the CBD. There is a dense, radial transport system free of congestion. Workers commute between residents and work places. Any other possible travel is neglected. The land is a featureless plain. All land parcels are identical and ready for residential use. This implies that the only spatial characteristic of each location in the city that matters to households is the distance from the CBD. We denote by $L(\omega)$ the land distribution in the city. It is further assumed that the land not occupied by households is used for agriculture, yielding a constant rent R_a. We assume that the city is closed in the sense that all the urban residents live in the city and there is no migration between the city and the outside world. We assume the absentee landownership. That is, absentee landlords own land.

1.3 The Alonso Model and its Extensions

A household is supposed to choose a residence that maximizes its utility subject to a budget constraint. The utility function $U(x,l)$ is assumed to be dependent on x (the amount of composite consumer good) and l (the lot size of the house). With the assumptions that $U(x,l)$ is continuous and increasing at all $x > 0$ and $l > 0$ and all indifference curves are strictly convex and smooth, we can describe the behavior of $U(x,l)$ as in Fig. 1.3.1.

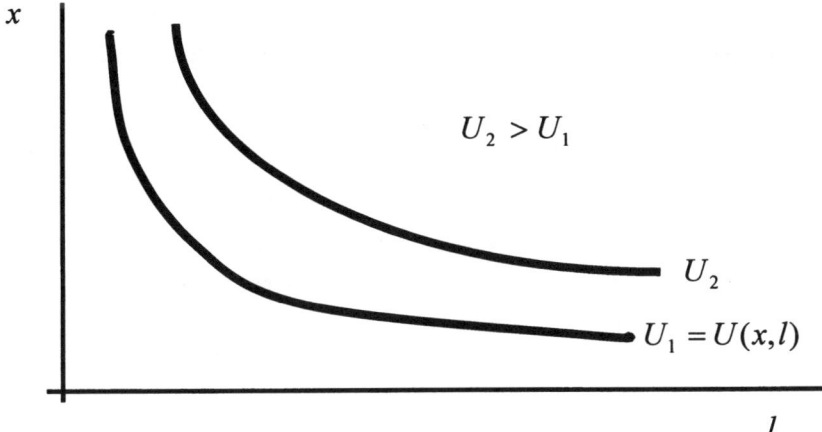

Fig. 1.3.1. Consumption and Indifference Curves

In the original Alonso model, the distance to the CBD is a variable that affects utility. The composite consumer good is selected as the numeraire. Let us consider that a household locates its house at distance ω (≥ 0) from the CBD. The budget constraint for this household is given by

$$x + R(\omega)l = Y - T(\omega) \tag{1.3.1}$$

where Y is the household's (exogenously) fixed income, $R(\omega)$ the rent per unity of land at location ω, and $T(\omega)$ the transport cost at ω. The household decides x, l and ω to maximize its utility. The residential choice of the household is to maximize $U(x,l)$ subject to (1.3.1), $l > 0$ and $x > 0$. It can be proved that (i) if $U(x,l)$ is continuous and increasing for $l > 0$ and $x > 0$; all indifferences are strictly convex and smooth and do not cut axes, and (ii) $T(\omega)$ is continuous and increasing at all $\omega \geq 0$, where $0 \leq T(0) < Y$ and $T(\infty) = \infty$, then the household's residential choice problem has a solution.

We use the bid rent approach to show the equilibrium of the Alonso model. The bid rent $\Psi(\omega,u)$ is the maximum rent per unit of land that the household can pay for living at ω, i.e.

$$\Psi(I,u) \equiv \max_{l,x}\left\{\frac{I-x}{l}\,\bigg|\, U(x,l) = u\right\}. \tag{1.3.2}$$

where $I(\omega) \equiv Y - T(\omega)$. It is shown that $\Psi(I,u)$ is continuously increasing in I and continuously decreasing in u (until $\Psi(I,u)$ becomes zero). When we solve the above maximization problem, we obtain the optimal lot size $l^*(I,u)$, which is called the bid-max lot size. It is shown that $l^*(I,u)$ is continuously decreasing in I and continuously increasing in u (until $l^*(I,u)$ becomes infinite). We depict the relation between $\Psi(I,u)$ and $l^*(I,u)$ as in Fig. 1.3.2. We see that the bid rent is given by the slope of the budget line at distance ω that is tangent to the indifference curve u.

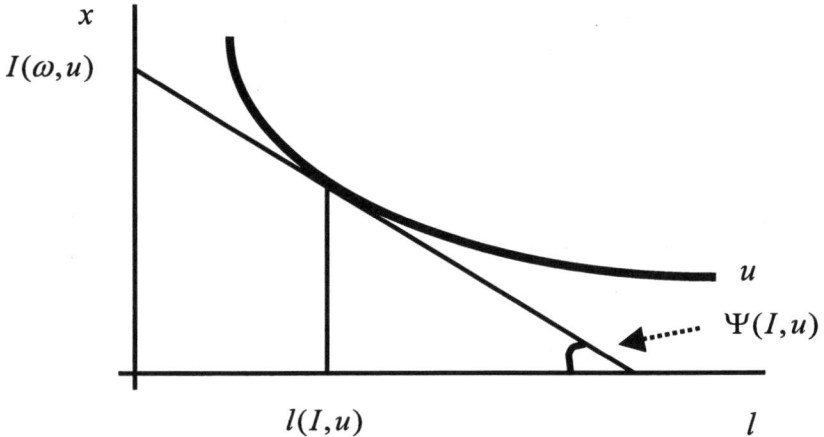

Fig.1.3.2. Bid Rent and Bid-max Lot Size

Given the market rent curve $R(\omega)$, we have that u^* is the equilibrium utility of the household and ω^* is an optimal location if and only if

$$R(\omega^*) = \Psi(\omega^*,u^*), \quad R(\omega) \geq \Psi(\omega,u^*), \quad \textit{for all } \omega. \tag{1.3.3}$$

The above equation provides the condition that an individual household chooses a location in the city. Every household treats the market rent curve as given and finds the most desirable location. We now consider behavior of all the households and landowners and determine the overall balance of demand and supply for land.

The competitive equilibrium land use refers to the situation in which equality between demand and supply for land is achieved everywhere. Since the households are identical, they must achieve the same maximum utility level independent of location in equilibrium. We denote by u^* the common maximum utility achieved in equilibrium. Since all land must be occupied either by housing or by agricultural use, we see that at each location the market land rent coincides with the maximum of the equilibrium bid rent and the agricultural rent, i.e.

$$R(\omega) = \max\{\Psi(Y - T(\omega), u^*), R_a\}, \quad \text{at each } \omega. \quad (1.3.4)$$

In other words, each location is occupied by the activity with the highest bid rent in equilibrium. Since $\Psi(Y - T(\omega), u^*)$ is decreasing in ω, the above equation becomes

$$R(\omega) = \begin{cases} \Psi(Y - T(\omega), u^*) & \text{for } \omega \le \omega_b, \\ R_a & \text{for } \omega > \omega_b \end{cases} \quad (1.3.5)$$

where ω_b is the urban fringe distance. Since land is fully occupied, we have

$$\begin{aligned} n(\omega) l^*(Y - T(\omega), u^*) &= L(\omega), \quad \omega \le \omega_b, \\ n(\omega) &= 0, \quad \omega > \omega_b \end{aligned} \quad (1.3.6)$$

where $n(\omega)$ is the household distribution at equilibrium. Since N households are located in the city, we have

$$\int_0^{\omega_b} \frac{L(\omega)}{l^*(Y - T(\omega), u^*)} d\omega = N. \quad (1.3.7)$$

The equilibrium values of $R(\omega)$, $n(\omega)$, $l(\omega)$, u^* and ω_b are determined by (1.3.5)-(1.3.7). It can be shown that the system has a unique equilibrium (Fujita, 1989, Arnott, 1996a).

This model provides a prototype of urban systems. It is partial in the sense that it takes the number of jobs at the CBD as given. There are many extensions of the model. Beckmann published a path-breaking paper in 1969. His paper incorporated many of the key assumptions in contemporary urban economics such as centralized CBD employment, a dense radial road system, and malleable housing. The paper introduced a distribution of income among households and explored how income and housing consumption vary with distance from the CBD. It is shown that if all households have an identical preference structure, then the wealthier households live further out and consume more space in equilibrium. Beckmann made another pioneering work in 1976. He considered a linear bounded landscape without a predetermined center (Beckmann, 1976). There is a single type of agents who derive utility from their interaction. Crowding which is a function of residential density has a negative impact on utility. At the initial state agents are distributed uniformly over the land and thus crowding is equal over space. On the other hand, it is assumed that those in central locations will enjoy higher degree of accessibility. It is shown that competition for land will eliminate this advantage through agglomeration. A bell-shaped population profile in equilibrium results from the trade-off between the propensity to interact and the aversion to crowding. Beckmann's model excludes the polycentric city as a possible outcome. As an extension of Solow and Vickrey (1971) and Beckmann (1976), Ogawa and Fujita (1980) introduced multiple types of interactions. Their model includes possible existence of multiple urban centers. Since then, there are many works on polycentric cities (Asami et a., 1990, Papageorgiou and Pines, 1999).

Some urban models have been developed within a general equilibrium framework. The general equilibrium models are concerned with the determination of the size and productive capacity of a monocentric city's employment, the socially efficient amount of land that must be allocated to road transportation around the center, the optimal degree of traffic congestion, and the quantity of housing that the market will produce (Mills, 1972, Dixit, 1973, Hardwick and Hardwick, 1974). The issues related to justification of equilibria in spatial economies are also raised (e.g., Stahl, 1985). The existence theorems are challenged when various spatial factors cause non-convex. For instance, Berlin and Tee Rae (1987) constructed examples in which all of the classical conditions, such as continuity and convexity of preferences hold but the utility of consumers explicitly depend on location. They demonstrated no existence of equilibrium in such models. Since each consumer is indivisible and unable to consume space at more than one location at a time, this generates a non-convexity in his consumption set. In urban economics the location-dependent utility may arise in many ordinary circumstances. For instance, it may arise either form the disutility of travel time to the CBD (Alonso, 1964), from traffic congestion (Solow, 1972), or industrial pollution (Diamond and Tolley, 1981), local public goods (Mills, 1987), and spatially distributed amenities (Kanemoto, 1980), or even from racial prejudice (Rose-Ackerman, 1975). The concentration of many people may be associated with different types of externalities such as traffic congestion, pollution, noises, racial discrimination, and neighboring amenities.

1.4 Imperfect Competition

It is well observed that firms selling similar products may either agglomerate or disperse in space. There are many reasons for these. For instance, the location of natural resources used as inputs, interaction externalities such as comparing shopping, and accessibility to customers may affect the location of firms in some way or another. There are different approaches to model interfirm competition over space. Hotelling (1929) gave the seminal work on spatial oligopoly. He studied the pattern of location of two sellers of a homogeneous product. Buyers are assumed to be evenly distributed on a linear market. Each buyer is assumed to consume one unit of output and bear all transport costs. By this assumption, complex of demand sides is ignored. For each pair of locations chosen by the firms, Hotelling calculated the equilibrium prices they would set and then introduced these equilibrium prices back into the firms' profit functions. He studied a subgame perfect equilibrium in a two-stage location-price game. He showed that the two salesmen will be located at the same place as a result of the competition and the equilibrium is stable. This conclusion also implies that competition between firms would lead them to provide essentially the same product. Stores would locate back-to-back and political parties would propose similar manifestos. This effect is now often referred to as "the principle of minimum differentiation". Smithies (1941) showed that if we introduce a non-zero elasticity of demand and non-zero transport costs the two competitors will separate. Eaton and Lipsey (1975) showed that if there are more than two sellers, they tend to disperse rather than agglomerate. Greenhut (1956) gave another way to improve the Hotelling model, integrating the Hotelling approach and the cost least approach within a single framework. Since there are many possibilities of transportation costs, demand functions and other spatial factors, it has become clear that the Hotelling model is only one of many possible location strategies even for two salesmen. It was pointed out by D'Aspremont, Gabszewicz and Thisse (1979) that under Hotelling's original specifications, location tendencies cannot be derived because the outcome of the price game is not well defined. They showed that the minimum differentiation would never hold in location-price models à la Hotelling since when firms locate coincidentally, profits are driven to zero by intense price competition. This proof of the non-existence of pure strategy price equilibrium in the Hotelling model has led many researchers to construct alternative models of price formation.

Hoover (1937a, 1937b) approached spatial competition in a different way. He allowed firms to discriminate in price across consumers. In this model, even when firms have access to the same production and transport technologies, they will never agglomerate. Agglomeration will entail zero profits. There are many models further developed along this line. Mills (1967) constructed an imperfect competition model in an urbanized economy where equilibria exist because individual firms face sufficiently inelastic demands. Due to spatial advantages, firms would restrict their size and operate in regions with increasing returns. This generates an industrial town with a single, monopolistic firm. The city center is in effect the location chosen by that firm and residential location is determined as in the standard Alonso-type models. There are models emphasizing different monopolistic pricing policies. Greenhut and Ohta (1972, 1975) analyzed monopoly output under alternative forms of spatial pricing. Ohta (1976) further examined issues related to efficiency of production under conditions of imperfect competition. Beckmann (1976) examined the spatial monopolist's optimal price, quantity, and market boundary, and compared consumer surplus, profits and social benefits under the spatial pricing policies. Benson (1980, 1984) examined Löschian competition and spatial competition with Chamberlinian tangencies. Eaton showed the nonuniqueness of equilibrium in the Löschian model. There are also studies that assume that firms behave as Cournot oligopolists and discriminate over space (Greenhut and Greenhut, 1975, Norman, 1981, Philips, 1983).

1.5 Spatial Agglomeration

We use terms such as production center, commercial center, service center, village, town, city, metropolitan area, and megapolitan area to describe concentration of human settlements and economic activities. Economic geography, urban and regional economics are concerned with the characteristics of these concentrations and their distribution over time and space. People interact with each other in various ways. They exchange tangibles and intangibles in the form of goods, services and ideas. People and their actions are concentrated because concentrations would bring about utility. Firms often prefer to locate in large cities because of proximity to other firms. According to Weber (1909:126), "An agglomerative factor ... is an 'advantage' or cheapening of production or marketing which results from the fact that production is carried on to some considerable extent at *one* place". He found out that agglomerative economies might arise, for instance, from simple enlargement of plant or from close local association of several plants.

There are agglomeration diseconomies such as congestion costs in urban formation. Congestion of firms and people would require high construction costs of building, oblige workers to commute long distance in crowded trains and raise land rent. Urban economists have identified some causes of urban agglomeration. These causes include diversity of the resource base, economies of scale in the provision of public sectors, and agglomerative economies of scale in private production. But earlier work on agglomeration tended to be descriptive (Cheshire and Evans, 1991). Most theoretical models of agglomeration are based on external scale economies and assume that the production function of a firm shifts out as city or industry size increases. External scale economies in cities are attributed to positive externalities that arise when firms locate near other firms. For instance, Henderson (1986) gave the following four factors that seem to capture the nature of these external effects: (1) economies of intraindustry specification where great industry size permits further specification among firms; (2) labor market economies where industry size reduces search costs and more effectively utilize human capital; (3) scale effects of communications where industry size economizes communications and speeds up adoption of new innovations, and (4) scale of providing public intermediate inputs such as infrastructures.

As in contemporary trade theory and growth theory, the Dixit-Stiglitz approach of monopolistic competition has exhibited important influences on contemporary development of spatial economics. The new trade theory applies models of imperfect competition drawn from the theory of industrial organization. It has made a major stimulus to the urban study. Abel-Rahman (1988) explained urban agglomeration on the basis of the Dixit-Stiglitz approach to determine equilibrium city size. The model uses the concept of product differentiation as an important factor in the formation of large cities. The supply side is characterized by monopolistic competition and decreasing average cost at the firm level. For the demand side, product variety is presented as a key factor in consumer agglomeration. Consumers buy differentiated products available in the local market. Consumer agglomeration in a particular area enlarges the equilibrium number of firms selling differentiated products, which raises utility in the Dixit-Stiglitz manner and at the same time augments the city's radius and raises rents at any given location. The urban equilibrium is achieved by the interactive forces of product diversity gains, consumer agglomeration, and changes in transport costs and rents. Krugman (1991) also applied ideas from the new trade theory to study spatial forms and cites. He tried to identify the conditions about when and why industry concentrates or disperse. He built a simple model of production and local increasing returns with transport costs. The model has multiple equilibria: one equilibrium exists with complete dispersion of activities and two others with spatial concentration in one region or the other. The model predicts that if one region has a slight advantage in production, then this slight advantage may generate extremely high density in one area and low density in others. Applying the frameworks used by Ethier (1982) in trade theory and Romer (1987) in growth theory (see also, Wang and Blomström, 1992, Fung and Ishikawa, 1992, Schmitz, 1989, Andersson and Zhang, 1990), Rivera-Batiz (1988) proposed a model that endogenizes agglomeration economies from both the production and consumption sides. It is assumed that an increase in city sizes enlarges the variety of consumer services locally available, shifting upwards household utility, and is associated with an agglomeration of industrial producers that raise the derived demand for local business, allowing increased specialization among them. The increased number of producer or business services available in the city improves the productivity of the industrial base of that city and results in endogenous external economies of scales.

1.6 Spatial Structures with Population and Knowledge

Ideas about increasing returns and economic structure and growth existed for a long time. Adam Smith emphasized increasing returns to scale to explain both specialization and economic growth. Marshall investigated the implications of increasing returns for economic development and externalities. The incompatibility of increasing returns and perfect competition has been recognized. It has become clear that it is difficult to analyze behavior of economic systems with non-constant returns to scale within traditional conceptual economic frameworks. There are also decreasing returns to scale in economic evolution. Malthus' theory of population provides one of possible factors for decreasing returns. Population is a key variable for the literature of non-constant returns to scale economies. Urbanization is often identified as a process that brings rural population into cities where industrial growth and expansion of the service sector are realized through economies of scale in production and increased division of labor. A U-shaped relationship between economic development and population concentration is often observed (Alperovich, 1992). In the initial stage of economic development population begins to concentrate in a small number of urban cores. Gradually, population disperses into suburbans as a result of increased land costs and deteriorating central environments. Subsequently, metropolitanization of the population becomes a predominant regional structure.

Marshall (1989) discussed the role of cities in spurring accumulation of human capital. Lucas (1988) emphasized the role of cities in analyzing interdependence between human capital, knowledge, and growth. Ideas diffuse quickly in cities. Unrelated agents come into contact and share ideas. Lucas analyzed the role of cities in facilitating the accumulation of knowledge spillovers in economic growth processes. High-tech industry is one of the fastest-growing sectors in developed economies. In high-tech regions, technological firms can enjoy the agglomeration benefits of a supportive local economy. Entrepreneurs, technicians and scientists accumulate human capital rapidly through being near the smarts. High-tech regions are often seen as leaders in national technological progresses and productivity improvements. In cities, creativity is not a linearly progressing phenomenon. Positive and negative effects of new ideas on different groups of people make it difficult to properly understand the mechanism of knowledge growth in cities. For instance, as more people are able to imitate an idea, more people can build on it. This would stimulate economic growth. On the other hand, as more people can imitate, there may be less incentive to innovate in the first place. Rauch (1991) examined human capital accumulation across cities, treating human capital spillovers to be local. In Rauch's approach, human capital externalities can be treated like any other urban amenity. He tried to use rent gradients to explain how living near educated people can enhance productivity. Since human capital in a city enhances everyone's productivity, wages in the city with high levels of human capital should be higher than those in other cities. Rauch introduced land rent as a balancing force for spatial equilibrium with identical population. He proved that in highly productive cities, rents should be high. Landowners are expected to charge more for the benefit from living near smart people.

1.7 City Systems

Urban and regional economists have proposed various reasons for spatial concentration of economic activities. These range from economies of scale internal to firms, proximity to markets and sources of raw materials, imperfect competition, social interaction, spatial business externalities, interindustrial linkages as well as consumers' taste heterogeneity. The history of city shows certain regularities in size distributions of cities and towns within national economies. Urban economists have tried to provide economic explanations for the existence of size distributions of cities. Lösch (1940) argued that there exists an optimal city size because urban utility is affected by the two opposite scale effects. The positive effect is due to the agglomeration of economic activities. The negative effect is due to crowding. Dixit (1973) and Henderson (1974) formally introduced the trade-off between these two scale effects and the consequent ∩ − shape of the utility level as a function of city size. In both models, urban production exhibits increasing returns to scale. The Dixit model assumes that residential areas exhibit scale diseconomies due to internalized traffic congestion, while the Henderson model takes account of the land scarcity and its effect on the urban production and housing. It is shown that there is an optimal population size if scale economies dominate for sufficiently small population size and scale diseconomies for sufficiently large.

Urban economics often treats individual cities in isolation. It was pointed out (e.g., Wheaton, 1979) that the conventional monocentric model of urban employment location is largely irrelevant to modern cities. In last few decades a large and increasing share of urban employment is concentrated in suburban centers. In other words, the standard Alonso's spatial structure should be extended to city systems with multiple centers. There are many models to analyze non-monocentric cities. Mills (1972) and Hartwick and Hartwick (1974) identified the conditions under which a city's production and residential sectors are either (1) segregated, with all production in the CBD and all residents in the suburbans, or (2) integrated, with firms interspersed with residents. Integration takes places when the cost of moving labor increases relative to the cost of moving output (to a central export node). Ogawa and Fujita (1980) suggested a model of nonmonocentric city in which integration occurs when the cost of moving labor increases relative to the cost of moving intermediate goods between firms. They showed that under certain conditions the city might have a number of employment subcenters. Helsley and Sullivan (1991) modeled non-monocentric cities, viewing a multinucleated city as a system of employment locations within a metropolitan area. They examined several non-spatial models of the allocation of a growing population to two production locations, where public capital must be installed prior to development of a production site. There are external scale economies in production and diseconomies in transportation. They tried to show how the development of multiple centers is affected by the fixed costs of public capital, differences in production technologies, and interactions between locations.

Alperovich (1982) analyzed the impact of agglomeration economies and diseconomies in production on the size distribution of cities in an urban hierarchy. Strong urbanization economies which are defined as intersectoral productivity effects may lead to the development of diversified large areas, whereas emerging localization economies which capture intrasectoral scale effects may foster specialized metropolitan areas if these economies occur in combination with possibilities for interarea trade. Moreover, the equilibrium size of metropolitan areas depends not only on the importance of localization and urbanization economies of scale within the areas themselves, but also on the costs related to the transfer of products and services between them. In a two-area model of single production factor (labor), Dierx (1990) studied how a change in intermetropolitan transfer costs may affect the dependency between two metropolitan areas and affect the relative size of both areas. The model incorporates both localization and urbanization economies of scale which are internal to one of the areas as well as scale economies due to the possibility of transfer of products between the two areas. It is shown that an improvement of interarea transport facilities may encourage trade and as a consequence decrease the disparity in area size.

1.8 Nonlinear Spatial Economic Dynamics

Spatial economies are in ceaseless creation and evolution. Spatial patterns of structural economies exhibit certain structural stability - they will last for some time. The relationship between economic forces and the spatial distribution of economic variables is centrally important for spatial economists. Formation and evolution of complicated dynamic structures is one of the most exciting areas of spatial economics. Urban dynamics are much more difficult to solve than they appear. Urban problems involve a large number of interacting components. It is not enough to understand how competitive and co-operative forces can shape spatial structures; we have to describe evolutionary processes. Economic mechanisms that lead to the rise and fall of cities are described by nonlinear dynamics which are characterized by the appearance of singularities and bifurcations. Structural changes occur when a balance between co-operative and competitive parameters passes through some critical points. As a consequence, an initially stable system becomes unstable and in a sequence of bifurcations a more complex time-dependent spatial structure appears. The fundamental bifurcations are steady state and Hopf bifurcations. In the former case, stationary modes branch off a basic state when the bifurcation parameter passes a critical point. In the latter case, cycles branch off a basic state. The type of bifurcations depends on the eigenvalues of the linearized equations. A zero eigenvalue implies a stationary solution whereas a pair of imaginary eigenvalues leads to cycles. Variation of a parameter may change the eigenvalue structure and thus lead to various bifurcations. In recent years, it has become evident that a large number of different structures and patterns can be formed far from equilibrium.

Urban growth and its dynamic processes have drawn a lot attention. The problem of explaining spatial economic evolution has attracted much interest and efforts for several distinct reasons. From the economic point of view, spatial economics is challenging because one has to develop a sophisticated theory to treat spaceless economics as a special case. Pattern formation in nonequilibrium systems constitutes a major branch of research in many scientific fields. The crucial question is why order may appear and which structures are selected among large manifolds of possibilities. Spatial pattern formation in economics should be an important component of this branch, not only because its mathematical structures are similar to other systems, for instance, in natural sciences, but also because it constitutes theoretically and practically important issues. From the mathematical point of view, spatial economic systems form nonlinear pattern-forming dynamics. The main characteristic feature of these systems is the existence of forces that drive them far from equilibrium. Such systems exhibit complicated behavior. In recent years, urban patterns formed by instabilities in evolutionary processes have received considerable attention (Anderson et al., 1988, Puu, 1989, Anas, 1992, Nijkamp et al., 1998, Rosser, 1991, Zhang, 1990b, 1991b, 1994d, Lorenz, 1993).

The traditional economic analysis could not handle with complicated issues related to urban evolution. It requires a rather sophisticated analysis to reveal workings of dynamic economic geography. It has been recently proposed that many systems with complex nonlinear dynamics when started in complex and random configurations, evolved towards a self-similar critical state. This situation is characterized by being very close to losing equilibrium, in the sense that small perturbations can lead to rearrangements of arbitrarily large sizes. Concepts such as catastrophe, bifurcation, chaos and fractals in nonlinear theory allow us to look at the complexity of economic reality in a new perspective and to consider irregularities as intrinsic entities. Many important theoretical questions remain unanswered. For instance, it is still an open question whether the economic dynamic complexity is a product of the randomness in the dynamic process or the result of deterministic instabilities.

1.9 The Purpose and Structure of the Book

The purpose of this book is to give a systematic exposition of urban economic theory. The emphasis is on theory. Each chapter is closely related to the other chapters, but is also self-contained. We are concerned with dynamic relations between spatial division of labor, division of consumption and the determination of prices structure in the spatial economy. We show dynamic interdependence between capital accumulation, knowledge creation and utilization, economic growth, price structures and spatial economic patterns. We will build our theory in a compact theoretical framework with a few concepts. The comparative advantage of our theory is that it uses only a few concepts and some simplified functional forms and accepts a few assumptions about behavior of consumers, producers and institutional structures, but it achieves rich conclusions and it is conceptually easy to extend and generalize the theory because of its consistency and simplicity.

We reviewed the literature in location theory and urban economics, even though this review is not comprehensive in the light of complex of the literature concerning with the subjects. This book is to develop spatial economic theory on the basis of the traditional theories. Although we will not strictly follow any special theory in our approach, we try to integrate the main forces of location and urban pattern formation in different theories within a compact framework. It should be remarked that this book does not claim to build a more 'general theory' than the traditional theories simply because this is technically an almost impossible task - each theory has been mathematically refined so much that a generalization of these refined and extremely complicated structures will lose some local structural characteristics. But this does not mean that our integrating work is not significant. On the contrary, we hold that the book is important since it shows a direction in which varied theories can be connected to each other in a consistent way.

Chapter 2 proposes a two-sector growth model with endogenous residential location in an isolated state. The model is a synthesis of the standard one-sector growth model, the Alonso location model with housing production and public land ownership. The model describes the dynamic interdependence between economic growth, transportation cost, housing demand and supply, residential location, wages and incomes over space. We show that the system has a unique stable equilibrium. We also examine the impact of changes in some parameters on the long-run growth and the economic geography.

Chapter 3 extends the model in Chapter 2, treating knowledge as an endogenous variable. We consider that knowledge creation and utilization is the driving force of modern economic development. This chapter proposes a compact framework to explain urban dynamics with endogenous urban pattern and capital and knowledge accumulation. We consider an economic system consisting of three sectors - industry, service, and university - in an isolated round island. The industrial sector produces commodities, which can be either consumed or invested. The service sector produces services. Services provided by, for example, restaurants, hotels, banking systems, travel agencies, transportation and communication systems, hospitals, and universities, are simultaneously consumed as they are produced. The university makes a contribution to knowledge growth. The university is a public sector in the sense that it is financially supported by the government's tax income. Accepting the Cobb-Douglas forms of production and utility functions, we can explicitly express the dynamics in term of two-dimensional differential equations for the total capital stocks and knowledge, with labor distribution, capital distribution, urban pattern, and knowledge at any point of time. It is shown that the system may have either unique or multiple equilibria and each equilibrium may be either stable or unstable, depending upon the combination of knowledge utilization efficiency and creativity of different activities. It is also shown how urban growth and urban pattern are affected by changes in knowledge accumulation and in scales of research and development activities in the long term.

Chapter 4 introduces family economics into spatial economic structural analysis. Urbanization is associated with dynamic interactions between family formation and structure, such as marriage and divorce, family size, child care, home production and non-home production and the time distribution of each family member, wealth distribution, consumption components and the mutual relationships between the family members. Over the years there have been a number of attempts to modify neoclassical consumer theory to deal with the economic issues related to family structure, working hours and the valuation of traveling time (Becker, 1965).

1.9 The Purpose and Structure of the Book

Nevertheless, only a few formal models are built to analyze spatial impact of sexual division of labor and consumption. The main purpose of this chapter is to propose a simple model on the basis of neoclassical growth theory, urban location theory, and family economics to examine the complexity of dynamic interaction between economic growth, residential location and the time distribution between work and leisure (time at home) of the male and female population.

In chapter 5 a two-group dynamic model with endogenous capital accumulation and residential location in an isolated linear system is proposed. The population is classified into two groups with different savings rates and human capital. We examine the dynamic interdependence among human capital, savings behavior, location choice and residential pattern. Our model is a synthesis, from a structural point of view, of the one-sector neoclassical growth model, the Kaldor-Pasinetti two-class model, the multi-group urban models, the Alonso location model, and the Muth housing model. We provide the conditions for existence of equilibria and stability of the dynamic system. Effects of changes in savings rates and human capital on long-run growth, income and wealth distribution between the two groups and residential structure are examined.

Chapter 6 introduces endogenous knowledge into the model with heterogeneous population proposed in Chapter 5. This chapter discusses whether economic development will result in the divergence or convergence of income and wealth distribution among various groups of people over space. The model is a synthesis of the Alonso model with multiple-income groups and the one-sector growth model with endogenous knowledge. It examines the dynamic interdependence between knowledge utilization, creativity, transportation conditions, savings behavior, location choice and residential patterns in a two-group economy with justice in the labor market. It is shown that the system may have a unique equilibrium or multiple equilibria and each equilibrium may exhibit stability or instability, depending upon the knowledge utilization and creation characteristics of the two groups. We will examine impact of changes in the population and the knowledge creation efficiency of the two groups upon long-run growth, wealth distribution and residential structure.

Chapter 7 proposes a spatial economic model to analyze a dynamic interaction between growth, preference change and spatial pattern formation in a perfectly competitive economic system. The model shows how some main ideas in growth theory and urban economics can be synthesized within a compact framework. In this model capital accumulation and preference change are the two forces for dynamics of economic geography. The conditions for existence of equilibria and stability are provided. We show that the economic geography may be either stable or unstable, depending on whether the households' propensity to save is weakened or strengthened when living conditions are improved. To show interdependence between capital, preference, and economic geography, we analyze the impact of changes in human capital on the system.

Chapter 8 proposes an equilibrium economic model of a one-city and one-farm spatial system. The model presumes the spatial division of labor as follows: farmers work and live in the agricultural area and workers are employed at the CBD and housed in the residential area. The problem is to determine spatial equilibrium of demand and supply and price structure with given territory, labor force, technology, preference and locational amenities. Effects of changes in urban amenity on division of labor and spatial prices are also provided.

Chapter 9 proposes an equilibrium model to explain differences in living costs between cities within a perfectly competitive framework. The economic system consists of two production centers. Each production centre specifies in supplying one commodity that is demanded by all the households. Services are city-specified in the sense that services are consumed only by the households who work in the city. We assume that all the households have an identical preference, professional changes and movement between the two cities are costless and the two cities have different levels of amenity and technology. It is proved that the perfectly competitive economic geography system has equilibria at which the land rent, service prices and wage rates are different between the two cities. We also examine the impact of changes in amenity upon the economic geography.

Chapter 10 is to propose a two-country growth model to investigate causes for the existence and persistence of trade patterns with different preferences, population and territory sizes between the two countries under conditions of perfect competition. Private consumption, residential location and savings are endogenously determined. The model may be considered as a synthesis of the standard neoclassical one-sector growth model, the neoclassical dynamic trade model with one commodity and the Alonso location model with public land ownership. It is shown that the dynamic system has a unique stable equilibrium. We examine effects of changes in marginal utilities of holding wealth and working efficiency difference on the equilibrium structure of international economies.

Chapter 11 provides an example of urban economic dynamics to show how one can explain urban complexity, while still working within the framework of traditional urban economic theory but by applying nonlinear dynamic methods. We are only concerned here with the dynamics of endogenous population growth and capital accumulation. For simplicity, the focus is first on an isolated island economy. Then, essential forces of various factors in forming urban pattern are addressed. Finally, the isolated economy is extended to an economic system consisting of interacting multiple islands. Issues related to urban oscillations and chaos are addressed.

Chapter 12 concludes this book, pointing out further possible extensions of this book and providing a broader vision of economic evolution than one actualized in this book.

2 Urban Growth with Housing and Spatial Structure

As mentioned in the introduction, numerous contributions to urban economics have followed the equilibrium theory of urban land market pioneered by Alonso (1964). This approach has been extended in many directions. However, it is argued that one of limitations of urban land-use models is that most of the models concentrate on the residential location and urban structure and neglect production aspects of urban dynamics. On the other hand, urban growth theory is another important theoretical approach to explain spatial phenomena. There are many studies of urban growth and dynamics (e.g., Richardson, 1973, Smith, 1975, Rabenau, 1979, Henderson, 1985, Fujita, et al, 2000). The development of this approach has been influenced by neoclassical growth theory and theory of international trade. From the point of view of economic geography, one of limitations of the urban growth models is that they omit urban land use. We see that the urban economics studies economic geography with wages and capital accumulation fixed and the urban growth theory concentrates on growth factors such as capital accumulation, omitting spatial structures of urban systems. The purpose of this chapter is to develop the dynamic urban growth model to describe the growth of an isolated economic system with economic geography. We explicitly deal with three important and connected aspects of urban economics: growth of production and housing sectors, capital accumulation, and distribution of population over time and space within a compact framework.

The remainder of this chapter is organized as follows. Section 2.1 defines the basic model. Section 2.2 shows how we can solve the dynamics with economic geography. Section 2.3 guarantees the existence of equilibrium and examines stability of the dynamic system. Sections 2.4 and 2.5 examine, respectively, the effects of changes in the population and the propensity to hold wealth on the long-run growth and economic geography. Section 2.6 concludes the chapter. The Appendix proves the conclusions in Section 2.2. This chapter is based on Zhang (1996d).

2.1 Urban Growth with Housing Production

We consider an isolated system. The population is homogenous. The households achieve the same utility level regardless of where they locate. All the markets are perfectly competitive. The system is geographically linear and consists of two parts -

22   2 Urban Growth with Housing and Spatial Structure

the CBD and the residential area. The isolated state consists of a finite strip of land extending from a fixed central business district (CBD) with constant unit width. We assume that all economic activities are concentrated in the CBD. The households occupy the residential area. We assume that the CBD is located at the left-side end of the linear territory, as illustrated in Fig. 2.1.1. As we can get similar conclusions if we locate the CBD at the center of the linear system, the special location will not essentially affect our discussion.

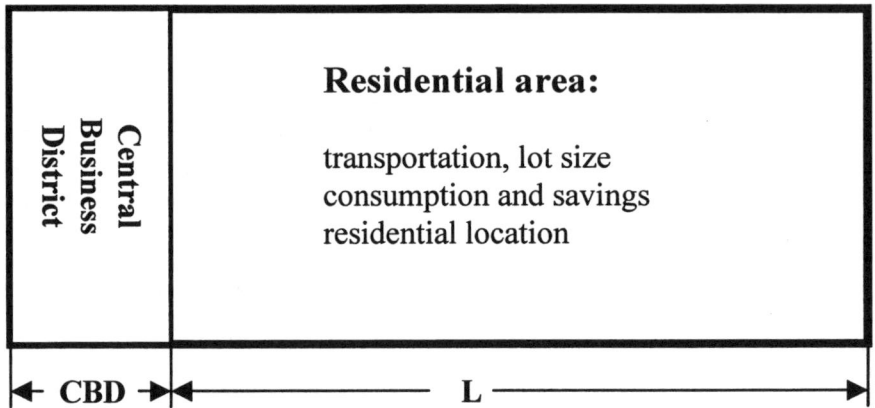

Fig.2.1.1. Economic Geography of the Isolated State

The system consists of two, industrial and housing, sectors. The industrial production is similar to that in the one-sector neoclassical growth model (Solow, 1956, Swan, 1956). We assume that the industrial product can be either invested or consumed. The housing production is similar to that in the Muth model (e.g., Muth, 1973). Housing is supplied with combination of capital and land. We assume that the total labor force is fully employed by the industrial sector. We select industrial good to serve as numeraire. As we assume that the transportation cost of workers to the city is dependent on the travel distance, land rent for housing should be spatially different. We assume that capital is freely mobile among the industrial and housing sectors under perfectly competitive environment.

We now describe the economic model of the isolated state. To describe the industrial sector, we introduce

N — the fixed population;

$K_i(t)$ — the capital stocks employed by the industrial sector at time t;

$w(t)$ and $r(t)$ — the wage rate and the rate of interest, respectively; and

2.1 Urban Growth with Housing Production

$F(t)$ and $C(t)$ — the output of the industrial sector and the total consumption of the commodity, respectively.

We assume that industrial production is carried out by combination of capital and labor force in the form of

$$F(t) = K_i^\alpha N^\beta, \quad \alpha + \beta = 1, \quad \alpha, \beta > 0 \tag{2.1.1}$$

where α and β are parameters. The marginal conditions for profit maximization are given by

$$r = \frac{\alpha F}{K_i}, \quad w = \frac{\beta F}{N}. \tag{2.1.2}$$

The optimal solution is illustrated as in Fig. 2.1.2.

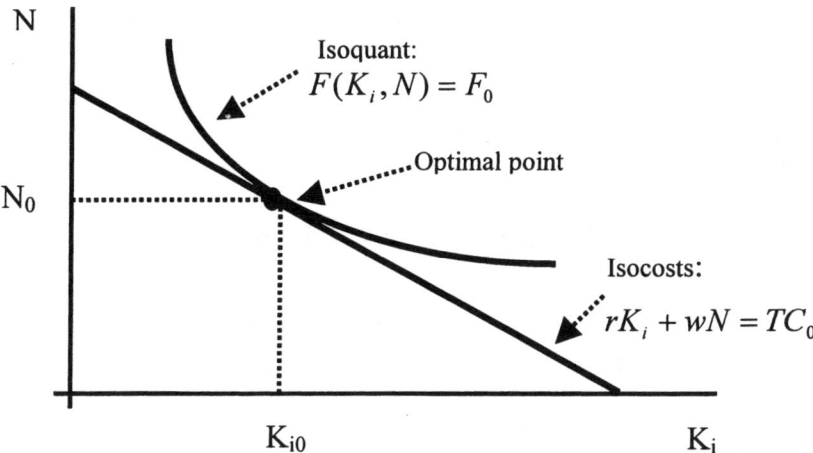

Fig.2.1.2. The Optimal Solution of Producers

We now describe housing production and behavior of households. First, we introduce:

L — the fixed (territory) length of the isolated state;
ω — the distance from the CBD to a point in the residential area;
$R(\omega,t)$ and $R_h(\omega,t)$ — land rent and housing rent per household at location ω;
$k(\omega,t)$ and $S(\omega,t)$ — capital stocks owned by and the savings out of the income made by the household at location ω, respectively;

$c(\omega,t)$ and $y(\omega,t)$ — the consumption and the net income of the household at location ω, respectively;

$n(\omega,t)$ and $L_h(\omega,t)$ — the residential density and the lot size of the household at location ω;

$K_h(t)$ — the capital stocks employed by the housing sector; and

$K(t)$ — the total capital stock of the economy.

Housing has a set of intrinsic properties which make it significantly from any other goods. Operation of housing market is varied in different regions and countries. There is a large literature on the complexity of housing dynamics (e.g., Muth, 1969, Hockman and Pines, 1980a, 1980b, Brueckner, 1981, Anas, 1982, Arnott, 1987), which explain various aspects of housing demand and supply. In this chapter, we limit ourselves to a simple case of housing technology and demand introduced by Muth. To explain space-dependent gradients for residential density and capital-land ratios, Muth introduces a commodity "housing" rather than land in describing dwelling conditions. Housing is produced with land and non-land inputs. Households have a derived demand for land, dependent on both preferences for housing and technical characteristics of housing production function. We follow this approach in explaining decision making of dwelling sites of households. The housing industry supplies housing services by combining land and capital. Let us denote $c_h(\omega,t)$ housing service received by the household at location ω. We specify the housing service production function as follows

$$c_h(\omega,t) = L_h^{\alpha_h}(\omega,t) k_h^{\beta_h}(\omega,t), \quad \alpha_h + \beta_h = 1, \quad \alpha_h, \beta_h > 0 \quad (2.1.3)$$

where $k_h(\omega,t)$ is the input level of capital per household at location ω. The marginal conditions are given by

$$r = \frac{\alpha_h c_h R_h}{k_h}, \quad R = \frac{\beta_h c_h R_h}{L_h}, \quad 0 \le \omega \le L. \quad (2.1.4)$$

According to the definitions of L_h and n, we have

$$n(\omega,t) = \frac{1}{L_h(\omega,t)}, \quad 0 \le \omega \le L. \quad (2.1.5)$$

The relationship between $k_h(\omega,t)$ and $K_h(t)$ is given by

2.1 Urban Growth with Housing Production

$$K_h(t) = \int_0^L n(\omega,t) k_h(\omega,t) d\omega. \tag{2.1.6}$$

To define net income, we now specify land ownership. For simplicity, we assume the public ownership, which means that the revenue from land is equally shared among the population. The total land revenue is given by

$$\overline{R}(t) = \int_0^L R(\omega,t) d\omega. \tag{2.1.7}$$

The income from land per household is given by

$$\overline{r}(t) = \frac{\overline{R}(t)}{N}.$$

The net income $y(\omega,t)$ per household at location ω consists of three parts: the wage income, the income from land ownership and the interest payment for the household's capital stocks. The net income is thus given by

$$y(\omega,t) = rk(\omega,t) + w(t) + \overline{r}(t). \tag{2.1.8}$$

Many previous models of residential location theory are developed with regards to rent theory since Alonso's seminal work (Alonso, 1964). In this approach residential location is modeled on the basis of the utility function. Location choice is closely related to the existence and quality of such physical environmental attributes as open space and noise pollution as well as social environmental quality. Basically following this approach, we assume that utility level U of the household at location ω is dependent on the temporary consumption level $c(\omega,t)$, housing conditions $c_h(\omega,t)$, the leisure time $T_h(\omega,t)$, the amenity $E(\omega,t)$, and the household's wealth $k(\omega,t) + S(\omega,t) - \delta_k k(\omega,t)$, where $\delta_k \,(\geq 0)$ is the fixed depreciation rate of capital, in the following way

$$U(\omega,t) = E T_h^\sigma c^\xi c_h^\eta (k + S - \delta_k k)^\lambda, \quad \sigma, \xi, \eta, \lambda > 0. \tag{2.1.9}$$

The assumption that utility is a function of wealth and consumption in the above form is used, for instance, in Zhang (1999, 2000). We will see that the above utility function provides a simple way to endogenously determine savings of the households.

We incorporate environmental quality into the consumer location decision. Distance from the CBD reflects two elements: the inconvenience of the distance and the value of the amenity of the surrounding area. The urban dynamics is influenced by many changing characteristics of environmental quality such as air quality, levels of noise pollution, open space, and other physical and social neighborhood qualities at each location. Environmental quality can be reflected in part by its effect on the location choice of the individual. Many kinds of externalities may actually exist at any location. Some may be historically given, such as historical buildings and climate; others such as noise and cleanness, may be endogenously determined by the location of residents. Therefore, amenity E at any location may be determined by many factors. In this chapter, we assume that households generally prefer a low-density residential area to a high one. As there tend to have more green, less noise, more cleanness and more safety in a low-density area, this assumption is acceptable. For simplicity of analysis, we specify $E(\omega,t)$ and $T_h(\omega,t)$ as follows

$$E(\omega) = \frac{\mu_1}{n(\omega)^\mu}, \quad T_h(\omega) = T_0 - \upsilon\omega, \quad \mu_1, \mu, \upsilon, T_0 > 0 \qquad (2.1.10)$$

The function $E(\omega,t)$ implies that the amenity level at location ω is determined by the residential density at the location. The function $T_h(\omega,t)$ means that the leisure time is equal to the total available time T_0 minus the traveling time $\upsilon\omega$ from the CBD to the dwelling site. Here, we neglect possible impact of congestion and other factors on the traveling time. Although it is easy to add traveling cost to the income budget, for simplicity of analysis, we neglect traveling cost. We assume that the amenity and traveling time are the main factors that affect the residential location. We illustrate the amenity and the leisure time respectively as in Figs. 2.1.3 and 2.1.4. It should be remarked that similar functional form of the residential externalities are widely used in the literature of urban economics (e.g., Kanemoto, 1980, 1996).

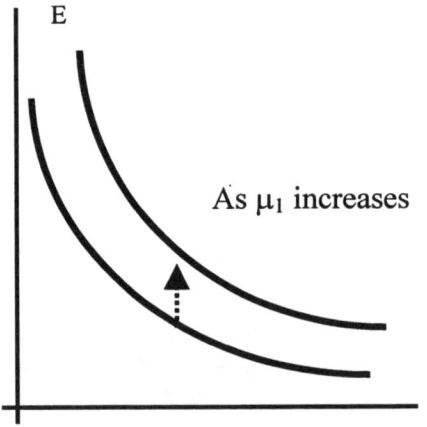

Fig.2.1.3. Amenity and Density

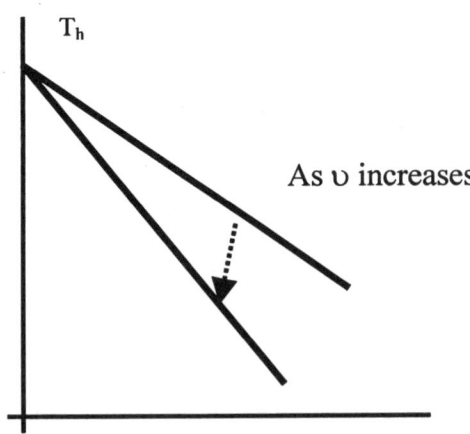

Fig.2.1.4. Leisure and Distance

2.1 Urban Growth with Housing Production

As the population is homogeneous, we will have

$$U(\omega_1, t) = U(\omega_2, t), \quad 0 \leq \omega_1, \omega_2 \leq L.$$

The budget constraint is given by

$$c(\omega,t) + R_h(\omega,t)c_h(\omega,t) + S(\omega,t) = y(\omega,t). \quad (2.1.11)$$

Maximizing U subject to the budget constraint yields

$$R_h(\omega)c_h(\omega) = \eta\rho y(\omega) + \eta\rho_0 k(\omega), \quad c(\omega) = \xi\rho y(\omega) + \xi\rho_0 k(\omega),$$
$$S(\omega) = \lambda\rho y(\omega) - (\xi + \eta)\rho_0 k(\omega) \quad (2.1.12)$$

where

$$\rho \equiv \frac{1}{\xi + \eta + \lambda}, \quad \rho_0 \equiv (1 - \delta_k)\rho.$$

The above equations mean that the housing consumption and consumption of the good are positively proportional to the net income and the wealth, and the savings is positively proportional to the net income but negatively proportional to the wealth.

According to the definition of $S(\omega,t)$, the capital accumulation for the household at location ω is given by

$$\frac{dk(\omega)}{dt} = S(\omega) - \delta_k k(\omega), \quad 0 \leq \omega \leq L.$$

Substituting $S(\omega,t)$ in (2.1.12) into the above equation yields

$$\frac{dk(\omega)}{dt} = sy(\omega) - \delta k(\omega), \quad 0 \leq \omega \leq L \quad (2.1.13)$$

where $s \equiv \lambda\rho$ and $\delta \equiv \delta_k + (\xi + \eta)\rho_0$. We see that the above capital accumulation equation is formally identical to the standard one-sector neoclassical growth model, even though the meanings of the savings rate s and the 'depreciation rate' δ are different from these in the traditional model.

As the state is isolated, the total population is distributed over the whole urban area. The population constraint is given by

$$\int_0^L n(\omega,t)\,d\omega = N. \qquad (2.1.14)$$

Similarly, the consumption constraint is given by

$$\int_0^L n(\omega,t)c(\omega,t)\,d\omega = C(t). \qquad (2.1.15)$$

We also have

$$\overline{S}(t) + C(t) = F(t) \qquad (2.1.16)$$

where $\overline{S}(t)$ is the total savings (= the total investment) of the economy, i.e.

$$\overline{S}(t) \equiv \int_0^L S(\omega,t)n(\omega,t)\,d\omega. \qquad (2.1.17)$$

The assumption that capital is always fully employed is given by:

$$K_i(t) + K_h(t) = K(t), \qquad (2.1.18)$$

$$\int_0^L k(\omega,t)n(\omega,t)\,d\omega = K(t). \qquad (2.1.19)$$

We have thus built the dynamic growth model with endogenous spatial distribution of wealth, consumption and population, capital accumulation and residential location. The system has 13 space-time-dependent variables, k, c, c_h, k_h, L_h, S, n, E, T_h, U, R_h, R, and y, and 10 time-dependent variables, F, K_i, K_h, K, , C, w, r, $\overline{S}$, $\overline{R}$, and $\overline{r}$. The system contains 23 independent equations. We now show that the problem has solutions.

2.2 The Dynamics in the Terms of $K(t)$

Before examining the dynamic properties of the system, we show that the dynamics can be described by the motion of a single variable $K(t)$.

Multiplying (2.1.13) by $n(\omega,t)$ and then integrating the resulted equation from 0 to L with respect to ω yields:

$$\frac{dK}{dt} = s\overline{Y}(t) - \delta K \tag{2.2.1}$$

where

$$\overline{Y}(t) \equiv \int_0^L y(\omega,t)n(\omega,t)\,d\omega . \tag{2.2.2}$$

We now show that $\overline{Y}(t)$ can be expressed as a function of $K(t)$.

Multiplying all the equations in (2.1.12) by $n(\omega,t)$ and then integrating the resulted equations from 0 to L with respect to ω, we obtain

$$\frac{rK_h}{\alpha_h} = \eta\rho\overline{Y} + \eta\rho_0 K, \quad C = \xi\rho\overline{Y} + \xi\rho_0 K,$$
$$\overline{S} = \lambda\rho\overline{Y} - (\xi + \eta)\rho_0 K \tag{2.2.3}$$

where we use $c_h R_h = rk_h/\alpha_h$ in (2.1.4). Substituting $\overline{S}$ and C in (2.2.3) into (2.1.16) yields

$$(\xi + \lambda)\rho\overline{Y} - \eta\rho_0 K = F . \tag{2.2.4}$$

From $r = \alpha F/K_i$ and $rK_h/\alpha_h = \eta\rho\overline{Y} + \eta\rho_0 K$, we have

$$\frac{\alpha K_h F}{\alpha_h K_i} = \eta\rho\overline{Y} + \eta\rho_0 K .$$

Substituting $\overline{Y} = (\eta\rho_0 K + F)/\rho(\xi + \lambda)$ obtained from (2.2.4) into the above equation yields

$$\Omega(K_i) \equiv \frac{\alpha K K_i^{-\beta}}{\alpha_h} - AK_i^\alpha - \frac{\eta \rho_0 K}{(\xi + \lambda)\rho N^\beta} = 0 \qquad (2.2.5)$$

where we use

$$K_h = K - K_i, \quad A \equiv \frac{\alpha}{\alpha_h} + \frac{\eta}{\xi + \lambda}.$$

We now show that for any given $K > 0$, $\Omega(K_i) = 0$ has a unique solution for $0 < K_i < K$. As $\Omega(0) > 0$, $\Omega(K) < 0$ and $d\Omega/dK_i < 0$, we see that $\Omega(K_i) = 0$ has a unique solution as a function of K. Let us represent this unique relationship by: $K_i(t) = \Lambda(K(t))$. From (2.2.5), we have

$$\frac{d\Lambda}{dK} = \frac{A\Lambda}{(\beta K/\alpha_h \Lambda + A)\alpha K} > 0. \qquad (2.2.6)$$

That is, an increase in the total capital stock will always increase the capital stocks employed in the industrial production. From $K_i = \Lambda(K)$ and $K_h = K - K_i$, we see that the capital stocks, K_i and K_h, employed by the industrial and housing sectors are uniquely determined as functions of the total capital stocks K at any point of time. From $K_h = K - K_i$ and (2.2.6), we have the impact of changes in K on K_h as follows

$$\alpha K \left(\frac{\beta K}{\alpha_h \Lambda} + A \right) \frac{dK_h}{dK} = \frac{\alpha \beta K^2}{\alpha_h \Lambda} + \alpha A K - A \Lambda$$

$$= \left(\frac{\alpha K}{\alpha_h \Lambda} - A \right) \beta K + A(K - \Lambda) > 0 \qquad (2.2.7)$$

where we use $K > \Lambda$ and $\alpha K/\alpha_h \Lambda - A = \eta \rho_0 K / \rho N^\beta \Lambda^\alpha (\xi + \lambda) > 0$ (from 3.5). That is, an increase in the total capital stock will always increase the capital stocks employed by the housing sector. From $F = \Lambda^\alpha N^\beta$, we see that F is a unique function of K. Substituting $\overline{Y} = \{\eta \rho_0 K + F(\Lambda)\}/\rho(\xi + \lambda)$ into (2.2.1), we have

$$\frac{dK}{dt} = s^*F\{\Lambda(K)\} - \delta^*K \tag{2.2.8}$$

where

$$s^* \equiv \frac{s}{(\xi + \lambda)\rho} > 0, \quad \delta^* = \delta - \eta\rho_0 s^* > 0.$$

At any point of time, the dynamic equation, (2.2.8), determines the value of the total capital stocks K. We can show that all the other variables are uniquely determined as a function of K and ω ($0 \leq \omega \leq L$) at any point of time. The following proposition is proved in the Appendix.

Proposition 2.2.1.
For any given (positive) level of the total capital stocks $K(t)$ at any point of time, all the other variables in the system are uniquely determined as functions of $K(t)$ and ω ($0 \leq \omega \leq L$). The dynamics of $K(t)$ is given by (2.2.8).

We can thus explicitly determine the motion of the system over time and space. It should be remarked that the result that the dynamics can be explicitly given in the simple form as (2.2.8) is important. This makes it possible to explicitly determine stability of the system.

2.3 Equilibrium and Stability

This section examines whether the system has a unique long-run equilibrium and whether the equilibrium, if exists, is stable. From (2.2.8), equilibrium is determined as a solution of the following equation:

$$s^*F\{\Lambda(K)\} = \delta^*K. \tag{2.3.1}$$

From (2.2.5), we have

$$K = \frac{A\Lambda^\alpha}{\alpha\Lambda^{-\beta}/\alpha_h - \eta\rho_0/\rho N^\beta(\xi + \lambda)}. \tag{2.3.2}$$

Substituting (2.3.2) into (2.3.1) yields

$$\Lambda = \left[\frac{\alpha}{\alpha_h \{\delta^* A / s^* + \eta \rho_0 / \rho(\xi + \lambda)\}}\right]^{1/\beta} N \qquad (2.3.3)$$

where we use $F = \Lambda^\alpha N^\beta$. Substituting (2.3.3) into (2.3.2), we directly determine K as a function of the parameters in the system. That is, the dynamic system has a unique equilibrium. From

$$\frac{d(s^* F - \delta^* K)}{dK} = \frac{d(s^* F)}{dK} - \delta^* = -\frac{\beta \delta^* K}{\beta K + \alpha_h A \Lambda} < 0 \qquad (2.3.4)$$

where we use $s^* F = \delta^* K$ and (2.2.6), we see that the unique equilibrium is stable.

We now determine equilibrium values of the other variables. The capital employed by the industrial sector is given by: $K_i = \Lambda$. From (2.1.1), (2.1.2) and (2.1.18), we directly get F, w, r and K_h. From (2.2.1), we have $\overline{Y} = \delta K / s$. Substituting this into (2.2.3), we have

$$C = \frac{\xi K}{\lambda}, \quad \overline{S} = \delta_k K. \qquad (2.3.5)$$

From the Appendix, we have

$$\overline{R} = \frac{\beta_h r K_h}{\alpha_h}, \quad \overline{r} = \frac{\overline{R}}{N}. \qquad (2.3.6)$$

From (2.1.13), we have

$$y = \frac{\delta k}{s}. \qquad (2.3.7)$$

Substituting (2.1.8) into this equation yields

$$k = \frac{sW}{\delta - sr} \qquad (2.3.8)$$

where

$$W \equiv w + \bar{r} = \frac{\beta - \beta_h + \alpha\beta_h K/K_i}{\alpha_h N} F \qquad (2.3.9)$$

in which we use (2.1.2). From $y = \delta k/s$ and (2.1.12), we obtain

$$c_h R_h = \frac{\eta k}{\lambda}, \quad c = \frac{\xi k}{\lambda}, \quad S = \delta_k k. \qquad (2.3.10)$$

From (2.1.4), we have $k_h = \alpha_h c_h R_h / r$. Substituting $c_h R_h$ in (2.3.10) into this equation yields

$$k_h = \frac{\alpha_h \eta R_h}{\lambda r}. \qquad (2.3.11)$$

Substituting (2.2.9) into $U(\omega)$ and then using $U(0) = U(\omega)$, we have

$$\frac{R_h(\omega)}{R_h(0)} = \left\{ \frac{E(\omega)T(\omega)^\sigma}{E(0)T(0)^\sigma} \right\}^{1/\eta}.$$

Substituting $c_h = L_h^{\alpha_h} k_h^{\beta_h}$ into $k_h = \alpha_h c_h R_h / r$, we have $R_h = rn^{\alpha_h} k_h^{\alpha_h} / \alpha_h$. Substituting this equation and (2.3.11) into $R_h(\omega)/R_h(0)$ in the above equation, we have

$$n(\omega) = n(0)\left(1 - \frac{\upsilon\omega}{T_0}\right)^B, \quad 0 \leq \omega \leq L \qquad (2.3.12)$$

where we use (2.1.10) and $B \equiv \sigma/(\alpha_h \eta + \mu) > 0$. From the above equation, we can analyze how the amenity parameter, the available time, the housing technology parameter, the traveling speed parameter and the propensity to use leisure time affect the residential density distribution. It can be seen that $R_h(\omega)$ and $n(\omega)$ are decreasing functions of the distance ω. We illustrate $n(\omega)$ in Fig. 2.3.1. Substituting (2.3.12) into (2.1.14), we have

$$n(0) = \frac{\upsilon N(1+B)}{\left[1 - (1 - \upsilon L/T_0)^{1+B}\right]T_0} \qquad (2.3.13)$$

where we use (2.3.8). We thus obtained $n(\omega)$. Here, we require $\upsilon L < T_0$. That is, the household's available time is more than that needed from traveling from the CBD to the boundary of the economic system.

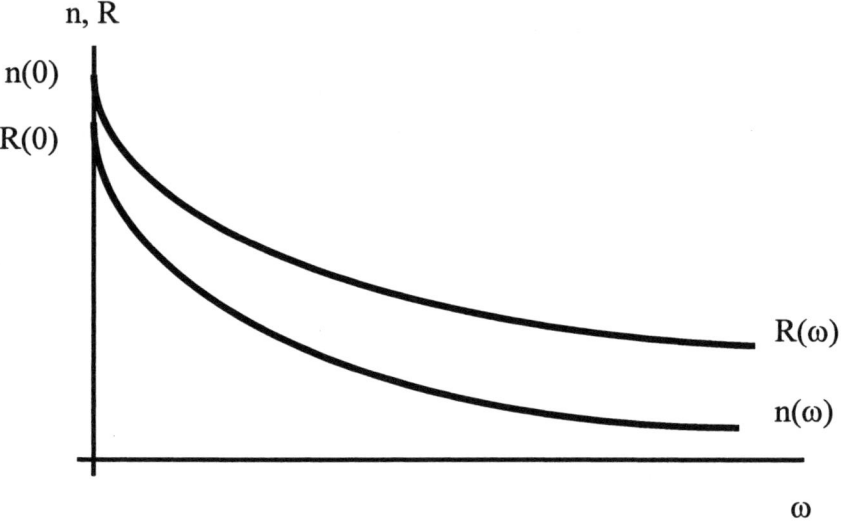

Fig.2.3.1. The Residential Density and Rent as Functions of the Distance

We get L_h, R_h, R and c_h, respectively, by (2.1.5), $R_h = rn^{\alpha_h} k_h^{\alpha_h} / \alpha_h$, (2.1.4) and (2.1.3). We have thus explicitly solved all the variables in the dynamic system with endogenous economic geography. Summarizing the above discussion, we have the following proposition.

Proposition 2.3.1.
The dynamic system has a unique stable equilibrium.

On the basis of (2.2.8), we can illustrate the unique equilibrium and its stability as in Fig. 2.3.2. It can be seen that the macrodynamic aspects of the model are quite similar to the Solow-Swan model. In the remainder of this study, we examine effects of changes in some parameters on the economic growth and geography.

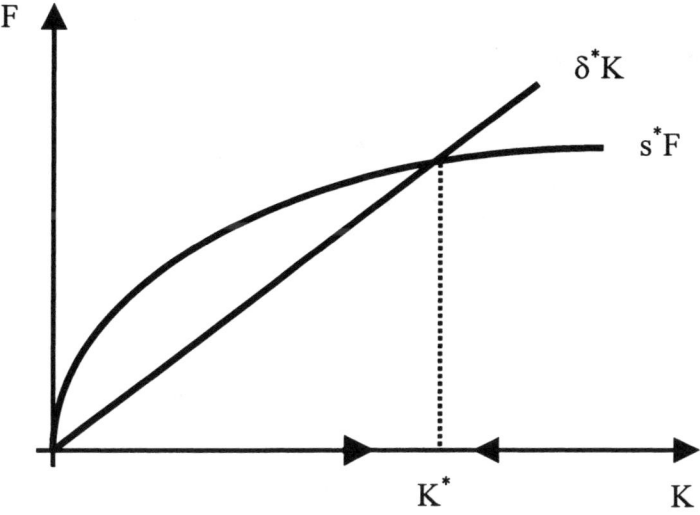

Fig. 2.3.2. Urban Growth and the Steady State

2.4 The Impact of the Population on Economic Geography

This section examines the impact of changes in the population on the long-run growth and the economic geography. First, taking derivatives of (2.3.3), (2.3.1) and $K_h = K - K_i$, we have

$$\frac{dK_i}{dN} = \frac{K_i}{N} > 0, \quad \frac{dK}{dN} = \frac{K}{N} > 0, \quad \frac{dK_h}{dN} = \frac{K_h}{N} > 0. \tag{2.4.1}$$

As the population is increased, the total capital stocks of the society and the capital stocks employed by the two sectors are increased. From (2.1.1) and (2.1.2), we directly have

$$\frac{dF}{dN} = \frac{F}{N} > 0, \quad \frac{dw}{dN} = \frac{dr}{dN} = 0. \tag{2.4.2}$$

The output of the industrial sector is increased and the wage rate and the rate of interest are not affected. It should be remarked that the neutrality of the wage and interest rates with regard to the population is due to the linearly proportional relationships between the capital stocks (K, K_i and K_h) and the population. We may describe the impact of change in N as in Fig. 2.4.1.

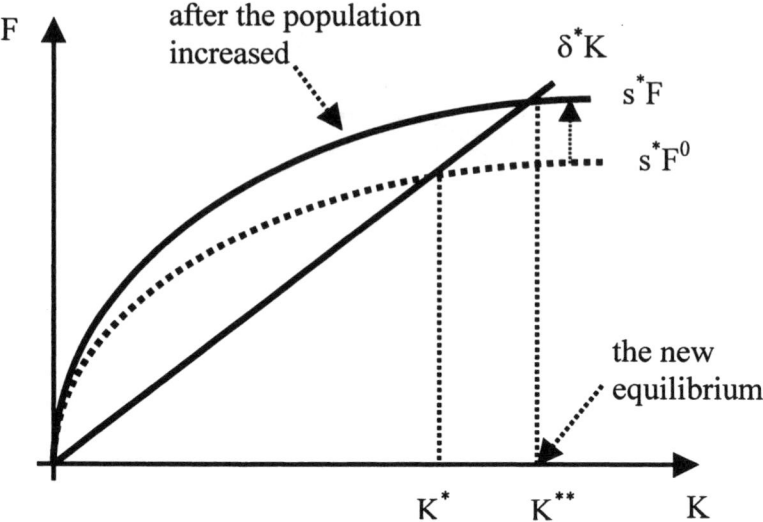

Fig. 2.4.1. The Impact of Population Growth

From (2.3.5), (2.3.6), (2.3.8), (2.3.9) and (2.3.10), we directly have

$$\frac{dC}{dN} > 0, \quad \frac{d\bar{S}}{dN} > 0, \quad \frac{d\bar{R}}{dN} > 0,$$

$$\frac{dk}{dN} = \frac{dc}{dN} = \frac{dS}{dN} = \frac{dk_h}{dN} = \frac{dy}{dN} = \frac{d\bar{r}}{dN} = 0. \quad (2.4.2)$$

We see that the capital k owned by per household, the consumption level per household c, the savings by per household S, the capital k_h utilized for per household, the net income per household, the revenue from land ownership per household are "neutral" to the population at any location. From (2.3.11) and (2.3.12), we have

$$\frac{1}{n(\omega)}\frac{dn(\omega)}{dN} = \frac{1}{N} > 0, \quad 0 \leq \omega \leq L. \quad (2.4.3)$$

From (2.1.5), $R_h = rn^{\alpha_h} k_h^{\alpha_h} / \alpha_h$, (2.1.4) and (2.1.3), we directly have:

$$\frac{dL_h(\omega)}{dN} < 0, \quad \frac{dR_h(\omega)}{dN} > 0, \quad \frac{dc_h(\omega)}{dN} < 0, \quad \frac{dR(\omega)}{dN} < 0. \quad (2.4.4)$$

We see that the housing market is affected by changes in the population.

2.5 The Propensity to Hold Wealth and the Equilibrium Structure

This section examines the impact of changes in the household's propensity λ to hold wealth on the urban equilibrium.

First, taking derivative of (2.3.3), (2.3.1) and $K_h = K - K_i$ with respect to λ, we have

$$\frac{dK_i}{d\lambda} = \frac{\alpha\xi/\alpha_h + \eta}{\alpha\alpha_h\lambda^2}\alpha F > 0, \quad \frac{dK}{d\lambda} = \frac{\alpha K}{K_i}\frac{dK_i}{d\lambda} + \frac{\xi K^2}{\lambda^2 F} > 0,$$

$$\frac{dK_h}{d\lambda} = \left(\frac{\alpha K}{K_i} - 1\right)\frac{dK_i}{d\lambda} + \frac{\xi K^2}{\lambda^2 F} \quad (2.5.1)$$

where we use

$$K = \frac{\lambda F}{(\lambda + \xi)\delta - (1 - \delta_k)\eta s}$$

which is obtained from (2.3.1). As the propensity to hold wealth is increased, the total capital and the capital employed by the industrial sector are increased; in the case of $\alpha K / K_i > 1$, the capital utilized by the housing sector is increased; but in the case of $\alpha K / K_i < 1$, the sign of $dK_h / d\lambda$ is uncertain. To examine the sign of $\alpha K / K_i - 1$, from (2.3.2) we get

$$\frac{\alpha K}{K_i} - 1 = \frac{\eta/(\xi + \lambda) - \beta/\alpha_h + \eta\rho_0 K_i^\beta / \alpha\rho N^\beta(\xi + \lambda)}{\alpha/\alpha_h - \eta\rho_0 K_i^\beta / \rho N^\beta(\xi + \lambda)}\alpha \quad (2.5.2)$$

where

$$\frac{\alpha}{\alpha_h} - \frac{\eta \rho_0 \Lambda^\beta}{(\xi + \lambda)\rho N^\beta} > 0.$$

Hence, the sign of $\alpha K / K_i - 1$ is the same as that of

$$\frac{\eta}{\xi + \lambda} - \frac{\beta}{\alpha_h} + \frac{\eta \rho_0 K_i^\beta}{(\xi + \lambda)\alpha \rho N^\beta} = A^*\left[\alpha_h \eta^2 + (1 - \delta_k)\alpha \eta \lambda \right.$$

$$\left. + (\alpha - \beta)(\xi + \delta_k \lambda)\eta - (\xi + \lambda)(\xi + \delta_k \lambda)\frac{\alpha \beta}{\alpha_h}\right] \quad (2.5.3)$$

where we use (2.3.3) to obtain K_i / N and

$$A^* \equiv \frac{1}{[(1 - \delta_k)\lambda \eta + \{(\xi + \lambda)\alpha / \alpha_h + \eta\}(\xi + \delta_k \lambda)]\alpha_h} > 0.$$

The sign of $dK_h / d\lambda$ is determined by the technology of industrial and housing production and the households' preference structure. As the housing demand and supply are dependent on the housing technology, preference structure and land value, this conclusion is intuitively acceptable. To explain how the new equilibrium value of K_h is achieved, we have to explain how all the equations in the system are affected by a shift in the preference parameter λ. We omit the explanation. As an illustration, we examine the case of $\alpha = \beta = \alpha_h = \beta_h = 0.5$. In this case, the sign of $dK_h / d\lambda$ is the same as that of

$$\eta^2 + (1 - \delta_k)\lambda \eta - (\xi + \delta_k \lambda)(\xi + \lambda).$$

If $\lambda = \xi = \eta$ then $dK_h / d\lambda < 0$; if δ_k is small and $\eta > \xi$, then $dK_h / d\lambda > 0$. From (2.1.1) and (2.1.2), we directly have

$$\frac{dF}{d\lambda} = \frac{\alpha F}{K_i}\frac{dK_i}{d\lambda} > 0, \quad \frac{dw}{d\lambda} = \frac{\alpha w}{K_i}\frac{dK_i}{d\lambda} > 0, \quad \frac{dr}{d\lambda} = -\frac{\beta r}{K_i}\frac{dK_i}{d\lambda} < 0.$$

$$(2.5.4)$$

The output of the industrial sector and the wage rate are increased and the rate of interest is reduced.

Taking derivatives of (2.3.5) with respect to λ yields

$$\frac{\lambda}{C}\frac{dC}{d\lambda} = \frac{\alpha\lambda}{K_i}\frac{dK_i}{d\lambda} + \frac{\xi K}{\lambda F} - 1, \quad \frac{d\overline{S}}{d\lambda} = \delta_k \frac{dK}{d\lambda} > 0,$$

$$\frac{d\overline{R}}{d\lambda} = \frac{\overline{R}}{r}\frac{dr}{d\lambda} + \frac{\overline{R}}{K_h}\frac{dK_h}{d\lambda}, \quad N\frac{d\overline{r}}{d\lambda} = \frac{d\overline{R}}{d\lambda}. \tag{2.5.5}$$

The total saving is increased and the total consumption of the commodity may be either increased or reduced. If the level of capital stocks employed by the housing sector is reduced, i.e., $dK_h/d\lambda < 0$, then the total revenue from land is certainly reduced. When $dK_h/d\lambda > 0$, the total revenue may be increased or decreased, depending on how much the rate of interest is reduced and the capital stocks of the housing sector is increased. From (2.3.8) and (2.3.9), we have

$$\frac{dk}{d\lambda} = \frac{s}{\delta - sr}\frac{dW}{d\lambda} + \frac{k}{\delta - sr}\frac{dr}{d\lambda} + \frac{(\xi + \eta)k}{(\delta_k \lambda + \xi + \eta - \lambda r)\lambda} \tag{2.5.6}$$

where

$$\frac{dW}{d\lambda} = \frac{W}{F}\frac{dF}{d\lambda} + \frac{\alpha\beta_h F}{\alpha_h N}\frac{d(K/K_i)}{d\lambda} \tag{2.5.7}$$

in which $dW/d\lambda > 0$ and $d(K/K_i)/d\lambda$ may be either positive or negative. It is not easy to explicitly judge the sign of $dr/d\lambda$. Taking derivatives of (2.3.10) with respect to λ, we have

$$\frac{dc}{d\lambda} = \frac{\xi}{\lambda}\frac{dk}{d\lambda} - \frac{c}{\lambda}, \quad \frac{dS}{d\lambda} = \delta_k \frac{dk}{d\lambda},$$

$$\frac{1}{k_h}\frac{dk_h}{d\lambda} = \frac{1}{k}\frac{dk}{d\lambda} - \frac{1}{\lambda} + \frac{1}{r}\frac{dr}{d\lambda}.$$

We can also analyze the impact of changes in λ on the other variables. As the expressions are too complicated, we omit the analysis.

2.6 On Extensions of the Basic Model

This chapter proposed a two-sector growth model of the isolated economy with endogenous residential location. The model is a synthesis of neoclassical growth theory and neoclassical urban economics. The system has a unique stable equilibrium. We also examined the impact of changes in the population and the propensity to save on the growth and the economic geography.

Numerous directions exist for extending the approach of this chaper. We may introduce more realistic representations of housing market dynamics and transportation systems with congestion (e.g., Arnott, 1979, Brueckner and Rabenau, 1981, Miyao, 1981). It is important to examine residential location with heterogeneous tastes (Wang, 1993).

Appendix

A.2.1 Proving Proposition 2.1

We already uniquely determined K_i, K_h and F as functions of K. The rate of the interest rate r and the wage rate w are uniquely determined by (2.1.2). From (2.1.4) and (2.1.5), we have $n(\omega)k_h(\omega) = \alpha_h R(\omega)/\beta_h r$. Substituting this into (2.1.6), we have $\overline{R} = \beta_h r K_h / \alpha_h$ where r and K_h are functions of K. We directly have: $\overline{r} = \overline{R}/N$. From (2.1.8), we see that $y(\omega)$ is a known function of K and $k(\omega)$ (as r, w and $\overline{r}$ are functions of K). We determine $k(\omega)$ as a function of K. We can get $c(\omega)$, $S(\omega)$ and $c_h(\omega)R_h(\omega)$ directly from (2.1.12). $\overline{S}$ and $k_h(\omega)$ are given by (2.1.16) and $k_h = \alpha_h c_h R_h / r$ from (2.1.4), respectively.

We now have five space-time dependent variables, L_h, n, c_h, R_h, and R, to determine. From (2.1.12), we see that c, $c_h R_h$ and $k + S - \delta_k k$ are known functions of K. Substituting (2.1.12) into $U(\omega)$ in (2.1.9), it is direct to see that we may have $U(\omega)$ in the form of

$$U(\omega, K) = \frac{f(\omega, K)}{R_h^\eta(\omega, K)}$$

where $f(\omega, K)$ is a function of ω and K. On the other hand, substituting $c_h = L_h^{\alpha_h} k_h^{\beta_h}$ into $r = \alpha_h R_h c_h / k_h$, we have $R_h = rn^{\alpha_h} k_h^{\alpha_h} / \alpha_h$ where r and k_h are given functions of ω and K. Substituting R_h into $U(\omega, K)$ and then using $U(\omega) = U(0)$, we have

$$n(\omega, K) = \left\{ \frac{f(\omega, K)}{f(0, K)} \right\}^{1/\alpha_h \eta} n(0, K).$$

Hence, if we can determine $n(0, K)$ as a function of K, then $n(\omega)$ is given. Substituting $n(\omega, K)$ in the above equation into the population constrain equation (2.1.14), we can explicitly get $n(0, K)$ as a function of K. From $L_h = 1/n$, $c_h = L_h^{\alpha_h} k_h^{\beta_h}$, $r = \alpha_h R_h c_h / k_h$ and $R = \beta_h R_h c_h / L_h$, we directly get L_h, c_h, R_h and R as functions of ω and K.

3 Spatial Pattern Formation with Capital and Knowledge

The previous chapter developed models based on the main ideas in the Alonso model and the one-sector neoclassical growth theory. We assumed that knowledge is exogenously given. However, when we study spatial economic evolution, it is too strict to assume knowledge as a given parameter in the dynamic process. Knowledge creation and utilization is the driving force of modern economic development. A main issue in economics is concerned with dynamic interactions between economic growth, knowledge creation and utilization. The literature on economic development has recently been centered on identifying different aspects of dynamic interactions between growth and knowledge accumulation. One of the first seminal attempts to render technical progress endogenous in growth models was initiated by Arrow (1962) emphasizing one aspect of knowledge accumulation - learning by doing. Uzawa (1965) introduced a sector specifying in creating knowledge into growth theory (see also, Zhang, 1990, 1999, Aghion and Howitt, 1992, 1998). The knowledge sector utilizes labor and the existing stock of knowledge to produce new knowledge, which enhances productivity of the production sector. Another approach was taken by, for instance, Kennedy (1964), Weisäcker (1966), Drandakis and Phelps (1966), and Samuelson (1965), who took account of the assumption of "inducement through the factor prices". Schultz (1981) emphasized the incentive effects of policy on investment in human capital. There are many other studies on endogenous technical progresses (e.g., Robson, 1980, Sato and Tsutsui, 1984, Nelson and Winter, 1982, Dosi, Pavitt and Soete, 1990, Johansson and Karlsson, 1990, Grossman and Helpman, 1991). In Romer (1986), knowledge is taken as an input in the production function and competitive equilibrium is rendered consistent with increasing aggregate returns owing to externalities. It is assumed that knowledge displays increasing marginal productivity but new knowledge is produced by investment in research, which exhibits diminishing returns. Various other issues related to innovation, diffusion of technology and behavior of economic agents under various institutions have been discussed in the literature. There are also many other models emphasizing different aspects, such as education, trade, R&D policies, entrepreneurship, division of labor, learning through trading, brain drain, economic geography, of dynamic interactions among economic structure, development and knowledge (e.g., Dollar, 1986, Krugman, 1991, Stokey, 1991, Zhang, 1991a, 1991c, 1997b, Barro and Sal-i-Martin, 1995). It may be argued that there is a theoretical limitation in those works. Capital and knowledge have not been integrated into a compact theoretical framework with economic structures and free markets. For instance, in the works by

Grossman and Helpman (1991) physical capital is almost totally neglected. This chapter proposes a compact framework to explain urban dynamics with endogenous urban pattern and capital and knowledge accumulation. The chapter is based on Zhang (1993a, 1993b).

3.1 The Urban Dynamics

We consider an economic system consisting of three sectors - industry, service, and university - in an isolated round island. There is neither migration nor trade with the outside world. The industrial sector produces commodities, which can be either consumed or invested. The service sector produces services. Services provided by, for example, restaurants, hotels, banking systems, travel agencies, transportation and communication systems, hospitals, and universities, are simultaneously consumed as they are produced. Services cannot be saved. The university makes a contribution to knowledge growth. The university is a public sector in the sense that it is financially supported by the government's tax income. Issues about public goods or bads, or neighborhood externalities are neglected. The system is geographically similar to that described by Alonso (1964). It is assumed that all land parcels are identical. Different from the previous chapter where we assumed the public ownership, this chapter assumes absentee ownership of the urban land.

The geography of the system consists of the city and the residential area. The city is monocentric; that is, it has a single pre-specified center of fixed size called the CBD (central business district). For simplicity of analysis, it is assumed that the city is a point. It is assumed that all economic activities and scientific research are carried out at the CBD. The residential area surrounds the CBD and occupied by households (that is, scientists and workers) working in the CBD. The distance from the CBD to a point in the residential area and the fixed radius of the island are denoted by ω and L, respectively. The system is illustrated as in Fig. 3.1.1.

Prices are measured in terms of the industrial good. It is assumed that the labor force is constant and that wage rates are identical among different professions because of perfect competition in the labor market. As it is assumed that the cost of transporting scientists and workers to the CBD is dependent on travel distance, land rent is spatially varied. We assume that capital is freely mobile among the three sectors by a perfectly competitive mechanism. This implies that the interest rate is identical in all parts of the system.

Subscripts i and s are used to denote the industrial sector and the service sector, respectively. We introduce

N — the fixed total labor force;
$K(t)$ — the total capital stock at time t;
$Z(t)$ — level of knowledge stock;

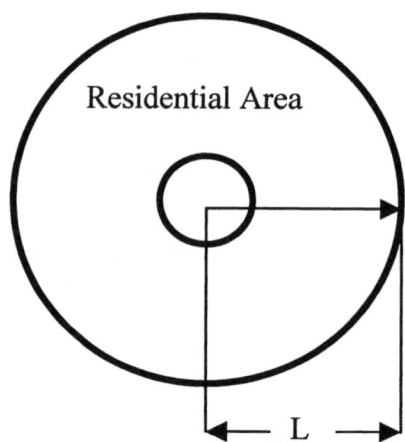

Fig.3.1.1. Economic Geography of the Monocentric State

$N_j(t)$ and $K_j(t)$ — labor force and capital stock employed by sector j, $j = i, s$, respectively;

$F_j(t)$ — the output level of sector j;

$N_r(t)$ and $K_r(t)$ — the number of scientists and capital stock employed by the university;

$p(t)$ — price level of services;

$w(t)$ and $r(t)$ — wage rate and rate of interest, respectively.

Production of the industrial sector
It is assumed that industrial production is carried out by combining knowledge, capital and labor force as follows

$$F_i(t) = Z^{m_i} K_i^{\alpha_i} N_i^{\beta_i}, \quad \alpha_i + \beta_i = 1, \quad \alpha_i, \beta_i > 0, \quad m_i \geq 0 \qquad (3.1.1)$$

where α_i, β_i and m_i are parameters.

Production of the service sector
We propose the production function of the service sector as follows

$$F_s(t) = Z^{m_s} K_s^{\alpha_s} N_s^{\beta_s}, \quad \alpha_s + \beta_s = 1, \quad \alpha_s, \beta_s > 0, \quad m_s \geq 0 \qquad (3.1.2)$$

where α_s, β_s and m_s are parameters.

3.1 The Urban Dynamics

As the product of the service sector is simultaneously consumed as it is produced, we have

$$C_s = F_s \tag{3.1.3}$$

where C_s is the consumption of services of the population.

Markets of capital and labor
Because perfect competition dominates the labor and capital markets, the wage rate is equal to the marginal product of labor and, the interest rate of capital is equal to the marginal product of capital

$$r = \frac{\tau^* \alpha_i F_i}{K_i} = \frac{\tau^* p \alpha_s F_s}{K_s}, \quad w = \frac{\tau^* \beta_i F_i}{N_i} = \frac{\tau^* p \beta_s F_s}{N_s} \tag{3.1.4}$$

in which $\tau^* \equiv 1 - \tau$ and τ is the tax rate on the output fixed by the government.

Savings and capital formation
The net income of the population $Y(t)$ is given by

$$Y(t) = F_i + pF_s. \tag{3.1.5}$$

Denote by s $(0 < s < 1)$ savings rate of the population out of the current gross national product. Capital accumulation is formulated by

$$\frac{dK}{dt} = sY - \delta_k K \tag{3.1.6}$$

in which δ_k is the given depreciation rate of capital.

Consumption choice
We assume that the level of utility U that households obtain from consuming industrial goods, services, and housing is expressed in the following form

$$U(\omega, t) = c_s^\gamma c_i^\xi L_h^\eta, \quad 1 > \gamma, \xi, \eta > 0 \tag{3.1.7}$$

where $c_s(\omega, t)$, $c_i(\omega, t)$ and $L_h(\omega, t)$ are consumption levels of service, industrial good and housing of a household at location ω and at time t. For simplicity, we require: $\gamma + \xi = \eta$. This hypothesis will simplify the analysis.

As the consumption budget is given by $(1-s)Y/N$, the consumer problem is defined by

$$\max U, \text{ s.t.: } c_i(\omega) + pc_s(\omega) + R(\omega)L_h(\omega) = \frac{(1-s)Y}{N} - \Gamma(\omega) \qquad (3.1.8)$$

where $\Gamma(\omega)$ is the total traveling cost between dwelling site ω and the CBD and $R(\omega)$ is land rent at location ω. We specify $\Gamma(\omega)$ by

$$\Gamma(\omega) = v_0 \omega^v, \quad v_0 > 0, \quad 0 < v < 1. \qquad (3.1.9)$$

We choose $v_0 = 1$. It is important to take account of possible influences of K and Z upon the transportation cost. For instance, as knowledge is improved, it is possible for $\Gamma(\omega)$ to decline or to increase. A reasonable way to relax this assumption on the transportation system is to introduce an endogenous transportation sector which takes part in economic competition.

The unique optimal solution is given by

$$c_i = \xi s_0 y, \quad c_s = \frac{\gamma s_0 y}{p}, \quad L_h = \frac{\eta s_0 y}{R} \qquad (3.1.10)$$

in which

$$s_0 \equiv \frac{1-s}{\gamma + \xi + \eta}, \quad y(\omega, t) \equiv s_1 Y(t) - \Gamma(\omega), \quad s_1 \equiv \frac{1-s}{N}.$$

We require $y(\omega, t) > 0$. This is guaranteed if $sY/N > L^v$, which simply implies that the total budget on consumption is more than the cost for any worker to travel from the CBD to the boundary of the island. In the remainder of this chapter it is assumed that this requirement is satisfied.

We denote by $n(\omega, t)$ the residential density at dwelling site ω. According to the definitions, we have

$$n(\omega, t) = \frac{1}{L_h(\omega, t)}, \quad 0 \leq \omega \leq L. \qquad (3.1.11)$$

The population is given by

$$2\pi \int_0^L n(\omega,t)\omega \, d\omega = N. \qquad (3.1.12)$$

The consumption constraints are given by

$$2\pi \int_0^L c_i(\omega,t)n(\omega,t)\omega \, d\omega = C_i(t),$$

$$2\pi \int_0^L c_s(\omega,t)n(\omega,t)\omega \, d\omega = C_s(t) \qquad (3.1.13)$$

where C_i is the consumption of industrial goods by the population.

Knowledge accumulation
We propose the following possible dynamics of knowledge

$$\frac{dZ}{dt} = \frac{\tau_i F_i}{Z^{\varepsilon_i}} + \frac{\tau_s F_s}{Z^{\varepsilon_s}} + \tau_r Z^{m_r} K_r^{\alpha_r} N_r^{\beta_r} - \delta_z Z \qquad (3.1.14)$$

where δ_z is a fixed depreciation rate of knowledge, τ_i, τ_s, τ_r, α_r, and β_r are non-negative parameters, and ε_i, ε_s, and m_r are parameters. We interpret $\tau_i F_i / Z^{\varepsilon_i}$ and $\tau_s F_s / Z^{\varepsilon_s}$ as the contribution to knowledge accumulation through the production sectors' learning by doing. In order to explain (3.1.14), we consider a case in which knowledge is a function of the total industrial output during a certain historical period

$$Z(t) = a_1 \{\int_0^t F_i(\theta) d\theta\}^{a_2} + a_3$$

in which a_1, a_2 and a_3 are positive parameters. The above equation implies that the knowledge accumulation through learning by doing exhibits decreasing (increasing) returns to scale in the case of $a_2 < (>) 1$. We interpret a_1 and a_3 as the measurements of the efficiency of learning by doing by the industrial sector. Taking the derivatives of the above equation yields

$$\frac{dZ}{dt} = \frac{\tau_i F_i}{Z^{\varepsilon_i}}$$

in which $\tau_i \equiv a_1 a_2$ and $\varepsilon_i \equiv 1-a_2$. In (3.1.14), the term $\tau_r Z^{m_r} K_r^{\alpha_r} N_r^{\beta_r}$ is called the contribution to knowledge growth by the university. This term is interpreted as that the knowledge production of the university is positively related to the capital stocks K_r employed by the university and the qualified labor input N_r of the university.

The budget of the university
Assuming that the financial resource of the university is from government's taxes upon the production sector, we have

$$\tau(F_i + pF_s) = rK_r + wN_r. \tag{3.1.15}$$

We now have to design a way to determine N_r and K_r. We assume that the government decides the number of scientists and the capital stock of the university in the following way

$$N_r = g_n N, \quad K_r = g_k K, \quad 0 < g_n, g_k < 1 \tag{3.1.16}$$

in which g_n and g_k are the policy variables fixed by the government.

Full employment of the labor force and capital
The assumption that the labor force and capital are fully employed is expressed by

$$N_i + N_s = (1 - g_n)N, \quad K_i + K_s = (1 - g_k)K. \tag{3.1.17}$$

We have thus finished building the model.

3.2 Equilibrium and Stability

As shown in the appendix, we can explicitly solve the variables, K_i, K_s, K_r, N_i, N_s, N_r, F_i, F_s, C_i, C_s, c_i, c_s, L_h, R, p, r, w, U, τ, and Y, as functions of K and Z at any location and at any point of time. The solutions are provided in the appendix. The urban structure, labor and capital distribution, and the monetary variables for any given values of the total capital $K(t)$ and knowledge $Z(t)$ at any point of time. Accordingly, the problem is to solve the dynamics of K and Z.

As the dynamic system is two-dimensional, we have to rewrite it in the term of two endogenous variables to analyze its behavior. It is now shown that the dynamics can

3.2 Equilibrium and Stability

be expressed by two differential equations consisting of the dynamics of K and Z. In Appendix A.3.1 we show that F_i, F_s and Y can be expressed as functions of K and Z in the following form

$$F_i(K,Z) = m_1 K^{\alpha_i} Z^{m_i}, \quad F_s(K,Z) = m_2 K^{\alpha_s} Z^{m_s},$$
$$Y(K,Z) = \frac{(\gamma + \xi) m_1 K^{\alpha_i} Z^{m_i}}{s\gamma + \xi} \quad (3.2.1)$$

in which

$$m_1 \equiv S_k^{\alpha_i}(S_n N)^{\beta_i}, \quad m_2 \equiv s_k^{\alpha_s}(s_n N)^{\beta_s}$$

where S_k, S_n, s_k, and s_n are parameters defined in (3.A.1.7) in Appendix A.3.1. Substituting (3.2.1) into (3.1.6) and (3.1.14) yields

$$\frac{dK}{dt} = \frac{(\gamma + \xi) F_i}{s\gamma + \xi} - \delta_k K,$$
$$\frac{dZ}{dt} = \frac{\tau_i F_i}{Z^{\varepsilon_i}} + \frac{\tau_s F_s}{Z^{\varepsilon_s}} + \tau_r n_0 Z^{m_r} K^{\alpha_r} - \delta_z Z \quad (3.2.2)$$

in which $n_0 \equiv g_k^{\alpha_r}(g_n N)^{\beta_r}$. The parameters, x_j, $j = 1, 2, 3$, defined by

$$x_1 \equiv \frac{m_i}{\beta_i} - \varepsilon_i - 1, \quad x_2 \equiv m_s - \varepsilon_s + \frac{\alpha_s m_i}{\beta_i} - 1,$$
$$x_3 \equiv m_r + \frac{\alpha_r m_i}{\beta_i} - 1 \quad (3.2.3)$$

are important in determining existence of equilibria and stability of the dynamic system. The following proposition is proved in Appendix 3.A.1.2. From the definitions of the parameters, it can be seen that x_j, $j = 1, 2, 3$, may be either positive or negative. We can show that any of 9 combinations of $x_j \geq 0$ or $x_k < 0$ is economically meaningful. We say that the contribution to knowledge accumulation by sector j exhibits increasing (decreasing) returns to scale in the system if $x_j > 0$ ($x_j < 0$).

Proposition 3.2.1.
We omit the case of $x_1 = x_2 = x_3 = 0$.

If $x_j \leq 0$ for all j, the system has a unique stable equilibrium;

If $x_j \geq 0$ for all j, the system has a unique unstable equilibrium;

In any of the remaining six combinations of $x_j \leq 0$ or $x_j \geq 0$ for all j, the system has either no equilibrium or two equilibria. When the system has two equilibria, the one with low values of K and Z is stable and the other one is unstable.

From the above interpretations of x_j, we may interpret the proposition as follows. If the two production sectors and university exhibit increasing (decreasing) returns in knowledge accumulation, the economy is unstable (stable). In the remainder of this chapter, we examine effects of changes in some parameters upon the system under presumed stability.

3.3 The Knowledge Accumulation Parameters

We take account of three sources for knowledge accumulation in (3.1.14). The parameter values, τ_i, τ_s, and τ_r, describe the efficiency of economic and scientific activities in knowledge creation. These parameters may be affected by institutions and by the work efficiency of scientists. As τ_i, τ_s, and τ_r have similar effects upon the system, it is sufficient for us to examine only one of them, for example, τ_r.

Taking derivatives of (3.A.1.10) and (3.A.2.1) with respect to τ_r yields

$$\Phi' \frac{dZ}{d\tau_r} = \frac{\Phi_3}{\tau_r} > 0, \quad \frac{dK}{d\tau_r} = \frac{m_i K}{\beta_i Z} \frac{d\Phi}{d\tau_r} > 0 \qquad (3.3.1)$$

in which $\Phi' = d\Phi/dZ$ (> 0 under presumed stability). An improvement in the efficiency of scientists' research will increase knowledge and capital stocks in the long term. It should be remarked that if the system is unstable, that is, $\Phi' < 0$, then $dZ/d\tau_r < 0$ and $dK/d\tau_r < 0$. The impact of changes in τ_r upon the system is sensitive to the stability conditions.

From (3.1.1)-(3.1.4) we have

3.3 The Knowledge Accumulation Parameters

$$\frac{dF_i}{d\tau_r} > 0, \quad \frac{dF_s}{d\tau_r} > 0, \quad \frac{dY}{d\tau_r} > 0, \quad \frac{dw}{d\tau_r} > 0.$$

An improvement in the efficiency of scientists' activities will result in increases in the output of the two economic sectors, the net income, and the wage rate. From $r = \alpha_i \tau^* F_i / K_i$, we have: $dr/d\tau_r = 0$. That is, shifts in τ_r have no effect upon the interest rate. As

$$\frac{Z}{p}\frac{dp}{d\tau_r} = \left(\frac{\beta_s m_i}{\beta_i} - m_s\right)\frac{dZ}{d\tau_r} \qquad (3.3.2)$$

in the case of $\beta_s m_i / \beta_i > m_s$ the price of services will be increased as the efficiency is improved.

We now examine effects of changes in τ_r upon spatial variables. Taking derivatives of (3.A.1.2) with respect to τ_r yields

$$\left[y(L) + \frac{\upsilon L^\upsilon}{2+\upsilon}\right]\frac{dR(L)}{d\tau_r} = s_1 R(L)\left[1 + \frac{2\upsilon L^\upsilon}{(2+\upsilon)y(L)}\right]\frac{dY}{d\tau_r} > 0.$$

(3.3.3)

We see that the land rent at the boundary is increased as the efficiency of research of scientists is improved. From (8.A.1), we have $R(\omega) = R(L)[y(\omega)/y(L)]^2$. From this equation we have

$$\frac{y(L)}{s_1 R(\omega)}\frac{dR(\omega)}{d\tau_r} = \left[\frac{\upsilon L^\upsilon}{\upsilon L^\upsilon + (2+\upsilon)y(L)} + \frac{s_1 Y + \omega^\upsilon - 2L^\upsilon}{y(\omega)}\right]\frac{dY}{d\tau_r}.$$

(3.3.4)

If the total consumption budget is more than twice of the cost for traveling from the CBD to the boundary, i.e., $s_1 Y > 2L^\upsilon$, then $dR(\omega)/d\tau_r > 0$. If the island is not large, it is acceptable to require $s_1 Y > 2L^\upsilon$ in the remainder of this chapter. As the research efficiency of scientists is improved, the land rent will be increased at any location in the long term.

We now examine the effects of changes in τ_r upon spatial distribution of consumption components. Taking derivatives (3.1.10) and (3.1.11), together with (3.3.2) and (3.3.4), we have

$$\frac{y(\omega)}{c_i(\omega)} \frac{dc_i(\omega)}{d\tau_r} = s_1 \frac{dY}{d\tau_r} > 0,$$

$$\frac{Z}{c_s(\omega)} \frac{dc_s(\omega)}{d\tau_r} = \left[\frac{m_i s_1}{\beta_i y(\omega)} - \frac{m_i \beta_s}{\beta_i} + m_s\right] \frac{dZ}{d\tau_r},$$

$$\frac{y(L)}{s_1 L_h(\omega)} \frac{dL_h(\omega)}{d\tau_r} = \left[\frac{L^\upsilon - \omega^\upsilon}{y(\omega)} - \frac{\upsilon L^\upsilon}{\upsilon L^\upsilon + (2 + \upsilon)y(L)}\right]\frac{dY}{d\tau_r},$$

$$\frac{1}{n(\omega)}\frac{dn(\omega)}{d\tau_r} = -\frac{1}{L_h(\omega)}\frac{dL_h(\omega)}{d\tau_r}. \qquad (3.3.5)$$

The effects upon consumption of services, housing and residential density are location-dependent. They may be either positive or negative, depending upon interactions of different factors. As there are so many variables and each variable is related to the others, to analyze the impact of changes in τ_r upon any variable we have to examine how all the variables and the functional relations among the variables are affected in the system.

3.4 The Impact of Government Intervention in Research

We now examine how the government can affect economic development by changing conditions of the research sector. From (3.A.1.10) and (3.A.2.1), we obtain effects of changes in g_n as follows

$$\Phi' \frac{dZ}{dg_n} = -\frac{\Phi_1}{1-g_n} - \frac{\Phi_2}{1-g_n} + \left(\frac{\beta_r}{g_n} - \frac{\alpha_r}{1-g_n}\right)\Phi_3,$$

$$\frac{1}{K}\frac{dK}{dg_n} = \frac{m_i}{\beta_i Z}\frac{dZ}{d\tau_r} - \frac{1}{1-g_n}. \qquad (3.4.1)$$

When β_r is small and α_r is large, an increase in the number of scientists tends to reduce the knowledge stock in the long term. When the percentage of scientists in the total labor force is already high, an expansion of university's scale tends to reduce

3.4 The Impact of Government Intervention in Research

knowledge in the long term. Even when β_r is large and g_n is small, if $\beta_r / g_n > \alpha_r /(1 - g_n) > 0$, an increase in g_n does not necessarily increase the level of knowledge stock. If the creative force, for example, of the industrial sector is strong, that is, Φ_1 is large, an increase in the number of scientists may not increase the knowledge stock in the long term.

As Φ_1, Φ_2, and Φ_3 respectively represent the creative forces of the industrial sector, the service sector and the university in equilibrium, we may interpret the case of $dZ/dg_n > 0$ as follows: when $\beta_r / g_n > \alpha_r /(1 - g_n) > 0$, if the creative forces of the industrial and service sectors are weak (i.e., Φ_1 and Φ_2 being small) and the creative force of the university is strong (i.e., Φ_3 being large), then an increase in the number of scientists will increase the knowledge stock in the long term. As dZ/dg_n is given by complicated interactional conditions, we can judge the sign of dZ/dg_n only under some simplified cases. As scientific research needs resources, we may conclude that effects of the number of scientists upon knowledge stock are not only related to the creativity of the scientists, but also are related to the creativity of the workers in the production sector. When the creativity of the workers is low and the number of the scientists is small, knowledge is increased if the number of scientists is increased.

We obtain the effects of changes in g_n upon output of the production sectors and income as follows

$$\frac{1}{F_i}\frac{dF_i}{dg_n} = \frac{m_i}{\beta_i Z}\frac{dZ}{dg_n} - \frac{1}{1-g_n},$$

$$\frac{Z}{F_s}\frac{dF_s}{dg_n} = \left(\frac{\alpha_s m_i}{\beta_i} + m_s\right)\frac{dZ}{dg_n} - \frac{Z}{1-g_n},$$

$$\frac{1}{Y}\frac{dY}{dg_n} = \frac{1}{F_i}\frac{dF_i}{dg_n}. \tag{3.4.2}$$

An increase of the number of scientists does not necessarily imply expanding production output. It is only when the expanded university scale greatly increases knowledge and the ratio of scientists to the total population is low that output of the industrial and service sectors and the total income may be increased in the long term. We see that the effects of government's research policy on the economic activities are dependent on the actual circumstances of the system.

The effects upon the wage rate, rate of interest, and price of services are given by

54   3 Spatial Pattern Formation with Capital and Knowledge

$$\frac{1}{w}\frac{dw}{dg_n} = \frac{1}{\tau^*}\frac{d\tau^*}{dg_n} + \frac{1}{F_i}\frac{dF_i}{dg_n},$$

$$\frac{1}{r}\frac{dr}{dg_n} = \frac{1}{\tau^*}\frac{d\tau^*}{dg_n} - \frac{2}{F_i}\frac{dF_i}{dg_n},$$

$$\frac{Z}{p}\frac{dp}{dg_n} = \left(\frac{m_i\beta_s}{\beta_i} - m_s\right)\frac{dZ}{dg_n} \qquad (3.4.3)$$

where

$$\frac{d\tau^*}{dg_n} = -\frac{(\gamma s + \xi)\beta\tau^{*2}}{(1-g_n)(\gamma+\xi)S_n} < 0.$$

We see that the wage rate is increased only when the industrial production is increased more than the tax rate. If industrial output is reduced, the wage rate is reduce by expanding the scale of the university. If output of the industrial production is expanded, the interest rate is reduced. But if the output of the industrial production is reduced, the interest rate may be either increased or decreased. In the case of $dZ/dg_n > 0$, the sign of dp/dg_n is the same as that of $m_i\beta_s/\beta_i - m_s$.

The effects upon the land rent are given by

$$\left[y(L) + \frac{\upsilon L^\upsilon}{2+\upsilon}\right]\frac{dR(L)}{dg_n} = s_1 R(L)\left[1 + \frac{2\upsilon L^\upsilon}{(2+\upsilon)y(L)}\right]\frac{dY}{dg_n},$$

$$\frac{y(L)}{s_1 R(\omega)}\frac{dR(\omega)}{dg_n} = \left[\frac{\upsilon L^\upsilon}{\upsilon L^\upsilon + (2+\upsilon)y(L)} + \frac{s_1 Y + \omega^\upsilon - 2L^\upsilon}{y(\omega)}\right]\frac{dY}{dg_n}.$$
(3.4.4)

As the scale of the university is expanded, the land rent may be either positive or negative. The sign of $dR(\omega)/dg_n$ is the same as that of dY/dg_n. We may also say that, if output of the industrial sector is reduced (increased), the land rent tends to decline (increase) at any location. The effects upon the consumption components at any location are given by

$$\frac{y(\omega)}{c_i(\omega)}\frac{dc_i(\omega)}{dg_n} = s_1 \frac{dY}{dg_n},$$

$$\frac{y(\omega)}{c_s(\omega)}\frac{dc_s(\omega)}{dg_n} = s_1 \frac{dY}{dg_n} - \frac{y(\omega)}{p}\frac{dp}{dg_n},$$

$$\frac{y(\omega)}{L_h(\omega)}\frac{dL_h(\omega)}{dg_n} = s_1 \frac{dY}{dg_n} - \frac{y(\omega)}{R(\omega)}\frac{dR(\omega)}{dg_n}. \tag{3.4.5}$$

We see that the effects upon the consumption components may be positive or negative, depending upon the whole structure of the system.

3.5 The Working Conditions of Scientists

In this section we examine effects of changes in the ratio g_k of the capital stock utilized by the university to the total capital stock of the economy. Taking the derivatives of (3.A.1.10) and (3.A.2.1) with respect to g_k yields

$$(1 - g_k)\Phi'\frac{dZ}{dg_k} = -\frac{\alpha_s \Phi_1}{\beta_i} - \frac{\alpha_s \Phi_2}{\beta_i} + \left(\frac{1 - \alpha_s - g_k}{g_k}\right)\alpha_r \Phi_3,$$

$$\frac{1}{K}\frac{dK}{dg_k} = \frac{m_i}{\beta_i Z}\frac{dZ}{dg_k} - \frac{\alpha_s}{(1 - g_k)\beta_i}. \tag{3.5.1}$$

If $1 < \alpha_s + g_k$, an increase in g_k will reduce knowledge and capital in the long term. When $1 > \alpha_s + g_k$, the sign of dZ/dg_n is dependent upon creative forces of the industrial sector, the service sector, and the university. Similarly to the previous section, we can examine the effects of changes in g_k upon the other variables.

3.6 On Knowledge Creation and Spatial Economic Evolution

This chapter proposed a model to analyze the interdependence among knowledge and capital accumulation and urban pattern formation. The model is essentially the combination of the Alonso model and neoclassical growth theory with endogenous knowledge. The model is constructed within a compact framework with perfect competition among all economic agents. In this dynamic system, capital and knowledge accumulation play an essential role in determining the long-run urban growth and urban pattern formation. Accepting specified Cobb-Douglas forms of production and utility functions, we can explicitly express the dynamics in term of two-dimensional differential equations for the total capital stocks and knowledge, with labor distribution, capital distribution, urban pattern, and knowledge at any point of time. It was shown that the system may have either unique or multiple equilibria and equilibria may be either stable or unstable, depending upon the combination of knowledge utilization efficiency and creativity of different activities. It was also shown how urban growth and urban pattern are affected by changes in knowledge accumulation and in scales of research and development activities in the long term.

Appendix

A.3.1 Expressing the Variables in Terms of K and Z

As we assume that the population is homogeneous, the utility level among households should be identical, i.e., $U(\omega,t) = U(L,t), 0 \leq \omega \leq L$. Utilizing this condition, (3.1.7) and (3.1.9), we have

$$\frac{y(\omega)}{R(\omega)} = \frac{y^2(L)}{y(\omega)R(L)} \qquad (3.A.1.1)$$

in which $y(L) = s_1 Y - L^\upsilon$ and $s_1 \equiv (1-s)/N$. Substituting

$$n(\omega) = \frac{1}{L_n(\omega)} = \frac{R(L)y(\omega)}{\eta s_0 y^2(L)}$$

into (3.1.12) yields

$$R(L) = \frac{(s_1 Y - L^\upsilon)^2 \eta s_0 N}{[s_1 Y - 2L^\upsilon/(2+\upsilon)]\pi L^2}. \qquad (3.A.1.2)$$

We see that (3.A.1.2) determines a unique relation between $R(L)$ and Y.

From (3.1.9), (3.1.13) and (3.2.1), we have

$$C_i = \xi G^*[Y(t)], \quad C_s = \frac{\gamma G^*[Y(t)]}{p(t)} \qquad (3.A.1.3)$$

in which

$$G^*[Y(t)] \equiv \frac{2\pi}{\eta} \int_0^L R\omega \, d\omega = \frac{\pi R_A}{y^2(L)} \left[(s_1 L Y)^2 - \frac{4 s_1 Y L^{2+\upsilon}}{2+\upsilon} + \frac{L^{2+2\upsilon}}{1+\upsilon} \right].$$
(3.A.1.4)

From $F_s = C_s$, $F_i = C_i + sY$, $Y = F_i + pF_s$ and (3.A.1.3), we get

$$G^*(Y) = \frac{(1-s)Y}{\gamma + \xi}.$$

Substituting this equation into (3.A.1.1) yields

$$C_i = \frac{(1-s)\xi Y}{\gamma + \xi}, \quad C_s = \frac{(1-s)\gamma Y}{(\gamma + \xi)p}. \tag{3.A.1.5}$$

By $F_s = (1-s)\gamma Y / p(\gamma + \xi)$ and $\alpha_i F_i / K_i = \alpha_s p F_s / K_s$, we have

$$\frac{K_s}{K_i} = \frac{(1-s)\alpha_s \gamma Y}{(\gamma + \xi)\alpha_i F_i}.$$

Substituting

$$F_i = C_i + sY = \left[\frac{(1-s)\xi}{\gamma + \xi} + s \right] Y$$

into the above equation yields

$$\frac{K_s}{K_i} = \frac{(1-s)\alpha_s \gamma}{(s\gamma + \xi)\alpha_i}. \tag{3.A.1.6}$$

From (3.1.4), we get $K_s / K_i = (\alpha_s \beta_i / \alpha_i \beta_s) N_s / N_i$. By this equation and (3.A.1.4), we get

$$\frac{N_s}{N_i} = \frac{(1-s)\beta_s \gamma}{(s\gamma + \xi)\beta_i}. \tag{3.A.1.7}$$

From (3.1.17), (3.A.1.6) and (3.A.1.7), we have

3 Spatial Pattern Formation with Capital and Knowledge

$$K_i = S_k K, \quad K_s = s_k K, \quad N_i = S_n N, \quad N_s = s_n N \qquad (3.A.1.8)$$

in which

$$S_k = \frac{(s\gamma + \xi)(1 - g_k)\alpha_i}{\alpha_i(s\gamma + \xi) + (1 - s)\alpha_s \gamma},$$

$$s_k = \frac{(s\gamma + \xi)(1 - g_k)\alpha_s \gamma}{\alpha_i(s\gamma + \xi) + (1 - s)\alpha_s \gamma},$$

$$S_n = \frac{(s\gamma + \xi)(1 - g_n)\beta_i}{\beta_i(s\gamma + \xi) + (1 - s)\beta_s \gamma},$$

$$s_n = \frac{(1 - s)(1 - g_n)\gamma\beta_s}{\beta_i(s\gamma + \xi) + (1 - s)\beta_s \gamma}. \qquad (3.A.1.9)$$

Substituting (3.A.1.8) into F_i, F_s, and Y, we have (3.2.1).

Utilizing

$$p = \frac{\beta_i N_s}{\beta_s N_i} \frac{F_i}{F_s} = \frac{(1 - s)\gamma}{s\gamma + \xi} \frac{F_i}{F_s},$$

$$wN_r = \frac{\tau^* g_n \beta_i N F_i}{N_i} = \frac{\tau^* g_n \beta_i F_i}{S_n},$$

$$rK_r = \frac{\tau^* g_k \alpha_i N F_i}{K_i} = \frac{\tau^* g_k \alpha_i F_i}{S_k},$$

we determine the tax rate τ by

$$\tau = \frac{wN_r + rK_r}{F_i + pF_s} = \frac{(s\gamma + \xi)}{\gamma + \xi}\left(\frac{\beta_i g_n}{S_n} + \frac{\alpha_i g_k}{S_k}\right)\tau^*.$$

That is

$$\frac{1}{\tau^*} = 1 + \frac{(s\gamma + \xi)}{\gamma + \xi}\left(\frac{\beta_i g_n}{S_n} + \frac{\alpha_i g_k}{S_k}\right). \qquad (3.A.1.10)$$

A.3.2 The Proof of Proposition 3.2.1

A balanced growth path of the isolated state is determined as a solution of the following equations

$$\frac{\gamma + \xi}{s\gamma + \xi} F_i = \delta_k K,$$

$$\frac{\tau_i F_i}{Z^{\varepsilon_i}} + \frac{\tau_s F_s}{Z^{\varepsilon_s}} + \tau_r n_0 Z^{m_r} K^{\alpha_r} = \delta_z Z. \qquad (3.A.2.1)$$

From the first equation in (3.A.2.1), we solve

$$K = DZ^{m_i/\beta_i}, \quad D \equiv \left[\frac{(\gamma + \xi)m_1}{(s\gamma + \xi)\delta_k} \right]^{1/\beta_i}. \qquad (3.A.2.2)$$

Substituting (3.A.2.2) into the second equation in (3.A.2.1) yields

$$\Phi(Z) \equiv \Phi_1(Z) + \Phi_2(Z) + \Phi_3(Z) - \delta_z = 0 \qquad (3.A.2.3)$$

in which

$$\Phi_1(Z) = \tau_i m_1 D^{\alpha_i} Z^{x_1}, \quad \Phi_2(Z) = \tau_s m_2 D^{\alpha_s} Z^{x_2},$$
$$\Phi_3(Z) = \tau_r n_0 D^{\alpha_r} Z^{x_3} \qquad (3.A.2.4)$$

We exclude the case of $x_1 = x_2 = x_3 = 0$. If $x_j \geq 0$, $j = 1, 2, 3$, we have

$$\Phi(0) < 0, \quad \Phi(\infty) > 0, \quad \Phi'(Z) = \frac{d\Phi(Z)}{dZ} > 0, \quad Z > 0.$$

The above properties of $\Phi(Z)$ guarantee that $\Phi(Z) = 0$ has a unique positive solution. Similarly, if $x_j \leq 0$, $j = 1, 2, 3$, we have

$$\Phi(0) > 0, \quad \Phi(\infty) < 0, \quad \Phi'(Z) = \frac{d\Phi(Z)}{dZ} < 0, \quad Z > 0.$$

We conclude that $\Phi(Z) = 0$ has a unique positive solution if all x_j are positive (negative) for $j = 1, 2, 3$, $\Phi(Z) = 0$ has a unique positive solution. For any of the remaining six ($= 2^4 - 2$) combinations of $x_j > 0$ or $x_j < 0$ ($j = 1, 2, 3$), the system has either two equilibria or none. We prove only one, $x_1 > 0$, $x_2 < 0$, and $x_3 < 0$, of these possible cases. The other cases can be similarly checked. It is easy to check that in the specified case, we have $\Phi(0) > 0$ and $\Phi(\infty) > 0$. This implies that $\Phi(Z) = 0$ has either no solution or multiple solutions. As

$$Z\Phi'(Z) = x_1\Phi_1 + x_2\Phi_2 + x_3\Phi_3 \qquad (3.\text{A}.2.5)$$

Φ' may be either positive or negative, depending on the parameter values. If $\Phi(Z) = 0$ has more than two solutions, there are at least two values of Z such that $\Phi' = 0$. Since $d(Z\Phi)/dZ > 0$ strictly holds for $Z > 0$, it is impossible for $\Phi' = 0$ to have more than one solution. A necessary and sufficient condition for the existence of two equilibria is that there exists a value of Z^* such that $\Phi(Z^*) < 0$ and $\Phi'(Z^*) = 0$.

The Jacobian at equilibrium is given by

$$J = \begin{pmatrix} -\beta_i \delta_k & m_i \delta_k K/Z \\ D_1 & D_2 \end{pmatrix} \qquad (3.\text{A}.2.6)$$

where

$$D_1 \equiv \frac{\alpha_i \tau_i F_i}{KZ^{\varepsilon_i}} + \frac{\alpha_s \tau_s F_s}{KZ^{\varepsilon_s}} + \alpha_r \tau_r n_0 Z^{m_r} K^{\alpha_r - 1},$$

$$D_2 \equiv \frac{(m_i - \varepsilon_i)\tau_i F_i}{Z^{\varepsilon_i + 1}} + \frac{(m_s - \varepsilon_s)\tau_s F_s}{Z^{\varepsilon_s + 1}} + m_r \tau_r n_0 Z^{m_r - 1} K^{\alpha_r} - \delta_z.$$

The two eigenvalues, ϕ_1 and ϕ_2, are given by

$$\phi^2 + (\beta_i \delta_k - D_2)\phi - \beta_i \delta_k D_2 - \frac{m_i \delta_k D_1 K}{Z} = 0. \qquad (3.\text{A}.2.7)$$

Analyzing the signs of ϕ_1 and ϕ_2, we have thus proved Proposition 3.2.1.

4 Urban Structure with Growth and Sexual Division of Labor

Three economic theories or modeling frameworks have played a significant role in the modern development of theoretical economics. They are the neoclassical growth theory (Burmeister and Dobell, 1970), the urban location theory (e.g., Alonso, 1964, Palma and Papageorgiou, 1988, Scotcmer and Thisse, 1992), and the family economics (Becker, 1976, 1981). Although the neoclassical growth theory has been applied and extended to explain issues about urban and regional growth and development (e.g., Henderson, 1985, Miyao, 1981), there are only a few theoretical efforts which introduce spatial aspects such as residential structures of cities and regions into the neoclassical growth theory. The previous chapters tried to synthesize the main ideas in the urban economics and neoclassical growth theory within a compact framework.

The main concern of family economics is the interactions between family formation and structure, such as marriage and divorce, family size, child care, home production and non-home production and the time distribution of each family member, wealth distribution, consumption components and the mutual relationships between the family members. Over the years there have been a number of attempts to modify the neoclassical consumer theory to deal with the economic issues related to family structure, working hours and the valuation of traveling time (Becker, 1965, 1985, Smith, 1977, Weiss and Willis, 1985, Persson and Jonung, 1997, 1998). There is an increasing amount of economic literature about, for instance, the sexual division of labor, marriage and divorce, and decision making about family size. It has been argued that increasing returns from human capital accumulation represent a powerful force creating a division of labor in the allocation of time between the male and female population. There are also studies of the relationship between home production and non-home production and the time distribution. Possible sexual discrimination in labor markets has also attracted much attention from economists. It should be noted that except a few studies (e.g., Beckmann, 1973), the sexual division of labor is not explicitly dealt with in urban economics.

The main purpose of this chapter is to propose a simple model on the basis of the neoclassical growth theory, the urban location theory, and the family economics to examine the complexity of dynamic interaction between economic growth, residential location and the time distribution between work and leisure (time at home) of the male and female population. As this chapter tries to synthesize the main ideas in the three approaches, the model is built as simply as possible at this initial stage.

4.1 Growth with Sexual Division of Labor and Location

As described in Fig.2.1.1, the system is geographically linear and consists of two parts - the CBD and the residential area. The isolated state consists of a finite strip of land extending from a fixed central business district (CBD) with constant unit width (Solow and Vickrey, 1971, Suh, 1988, Wang, 1993). It is assumed that all economic activities are concentrated in the CBD. The households achieve the same utility level regardless of where they locate. The residential area is occupied by households. It is assumed that the CBD is located at the left-hand side of the linear territory. As we can gain similar conclusions if the CBD is located at the center of the linear system, the special location does not essentially affect the discussion below.

Economic production consists of a single industrial sector. The industrial product can be either invested or consumed. Full employment of capital and the labor force are assumed. The industrial commodity is selected to serve as the numeraire. It is assumed that there is no sexual discrimination in the labor market. Here, by justice in the labor market we mean that any worker is paid according to the marginal value of 'qualified labor'. Markets are characterized by perfect competition. The total working times (to be defined) of the male and female populations are denoted by $N_1(t)$ and $N_2(t)$, respectively. The total qualified labor force $N^*(t)$ is defined by

$$N^* = zN_1 + N_2. \tag{4.1.1}$$

In (4.1.1), the parameter z simply means the difference in efficiency between the male and female populations. If $z > 1$, $z = 1$, $z < 1$, the male population's productivity per unit of time is higher than, equal to, or lower than that of the female population in respect to the industrial production. For simplicity, it is assumed that people of the same sex have the same level of productivity, although the productivity of one sex may be different from that of the other sex. Here, it is assumed that z is constant.

The production function of the economy is specified as follows

$$F = K^\alpha N^{*\beta}, \quad \alpha + \beta = 1, \quad \alpha, \beta > 0 \tag{4.1.2}$$

where $F(t)$ is the total output, $K(t)$ denotes the capital stocks employed by the industrial sector, and α and β are parameters.

4.1 Growth with Sexual Division of Labor and Location

The industrial sector maximizes its profit, $F - rK - w_1 N_1 - w_2 N_2$ where r is the rate of interest and w_1 and w_2 are the wage rates per unit of working time of the male and female populations respectively. The marginal conditions for the industrial sector are given by

$$r = \frac{\alpha F}{K}, \quad w_1 = \frac{\beta z F}{N^*}, \quad w_2 = \frac{\beta F}{N^*}. \tag{4.1.3}$$

To describe the behavior of the residents, the following parameters and variables are introduced

L — the fixed (territorial) length of the isolated state;
ω — the distance from the CBD to a point in the residential area;
$R(\omega,t)$ — the land rent at location ω;
$k(\omega,t)$ and $S(\omega,t)$ — the capital stocks owned by and the savings made by the family at location ω, respectively;
$c(\omega,t)$ and $y(\omega,t)$ — the consumption and the net income of the family at location ω, respectively; and
$n(\omega,t)$ and $L_h(\omega,t)$ — the residential density and the lot size of the family at location ω, respectively.

According to the definitions of L_h and n, we have

$$n(\omega,t) = \frac{1}{L_h(\omega,t)}, \quad 0 \leq \omega \leq L. \tag{4.1.4}$$

It is assumed that the total working time of each family may be different, depending on their residential location and the family's wealth. Let $T_1(\omega,t)$ and $T_2(\omega,t)$ denote, respectively, the working time of a husband and wife at location ω. According to the definitions of $R(\omega,t)$ and $T_j(\omega,t)$, the total working time $N_j(t)$ of sex j is given by

$$N_j(t) = \int_0^L T_j(\omega,t) n(\omega,t) \, d\omega. \tag{4.1.5}$$

That is, the total working time of each sex at any point of time is the sum of working time of the sex's population over space. Here, we omit any other possible impact of working time on productivity. For instance, if over-working reduces productivity per unity of working time, it is much more complicated to model the qualified labor force and the working time.

To define net income, it is necessary to specify land ownership. For simplicity, public ownership is assumed. This implies that the revenue from land is equally shared among the population. The total land revenue is given by

$$\overline{R}(t) = \int_0^L R(\omega,t)\,d\omega. \tag{4.1.6}$$

Each family's income from land ownership is given by

$$\overline{r}(t) = \frac{\overline{R}(t)}{N}.$$

The net income $y(\omega,t)$ of the family at location ω consists of three parts: its wage incomes, $T_1(\omega)w_1 + T_2(\omega)w_2$, the income $\overline{r}$ from land ownership, and the interest payment $rk(\omega)$ for its capital stocks. The net income is thus given by

$$y(\omega,t) = rk(\omega) + T_1(\omega,t)w_1(t) + T_2(\omega,t)w_2(t) + \overline{r}(t). \tag{4.1.7}$$

It is assumed that there is an aggregated utility function for each family. As each member of a family has his/her own utility function, the way in which the family game (for instance, the distribution of consumption goods among the family members) is actually played is very complicated. From microeconomic points of view, the family's behavior should be analyzed on the basis of all members' rational decisions. The "collective" utility function should be analyzed within a framework which explicitly takes account of the interactions between family's members (e.g., Becker, 1976, Heckman and Macurdy, 1980, Chiappori, 1988, 1992). Here, for simplicity of analysis, issues concerning possible conflicts and inequalities between family members are neglected.

It is assumed that the utility level U of a typical family at location ω is dependent on its temporary consumption level $c(\omega,t)$, lot size $L_h(\omega,t)$, the husband's leisure time $T_{h1}(\omega,t)$, wife's leisure time $T_{h2}(\omega,t)$, locational amenities $A(\omega,t)$, and the family's wealth $k(\omega,t) + S(\omega,t) - \delta_k k(\omega,t)$, where δ_k ($\delta_k \geq 0$) is the fixed depreciation rate of capital. For simplicity, we specify $U(\omega,t)$ as follows

4.1 Growth with Sexual Division of Labor and Location

$$U(\omega,t) = AT_{h1}^{\sigma_1}T_{h2}^{\sigma_2}c^{\xi}L_h^{\eta}(k + S - \delta_k k)^{\lambda}, \quad \sigma_1, \sigma_2, \xi, \eta, \lambda > 0.$$
(4.1.8)

It is necessary to specify a functional form for the locational amenities $A(\omega,t)$. In this study, it is assumed that households generally prefer a low-density residential area to a high density one. As it tends to be greater, quieter, cleaner and safer in low-density areas, this assumption is quite acceptable. We specify $A(\omega,t)$ as follows

$$A(\omega,t) = \frac{\mu_1}{n^{\mu}(\omega,t)}, \quad \mu_1, \mu > 0$$
(4.1.9)

where μ_1 and μ are parameters. As shown below, it is analytically not necessary to require μ to be positive. It should be noted that there are various ways to take account of other factors in defining the locational amenities function (e.g., Kanemoto, 1980).

Let T_0 denote the total available time (which is assumed to be equal between the two sexes). The time constraint requires that the amounts of time allocated to each specific use add up to the time available

$$T_j(\omega,t) + T_{hj}(\omega,t) + \upsilon\omega = T_0, \quad j = 1, 2$$
(4.1.10)

in which υ is a traveling time per unit of distance. The term $\upsilon\omega$ in (4.1.10) is equal to the traveling time from the CBD to the dwelling site ω. Here, possible impacts of congestion and other factors on the traveling time are neglected.

As the population is homogeneous, the utility level is identical throughout the state, irrespective of location, i.e.

$$U(\omega_1,t) = U(\omega_2,t), \quad 0 \le \omega_1, \omega_2 \le L.$$
(4.1.11)

Each family makes decisions about six variables, T_1, T_2, L_h, S, c, and ω, at any point of time. The financial budget is given by

$$c(\omega,t) + R(\omega,t)L_h(\omega,t) + S(\omega,t) = y(\omega,t).$$
(4.1.12)

Substituting (4.1.7) and (4.1.10) into (4.1.12), we rewrite the budget constraint as follows

$$c + RL_h + S + T_{h1}w_1 + T_{h2}w_2 = \bar{r} + rk + (T_0 - \upsilon\omega)(w_1 + w_2).$$
(4.1.13)

It is assumed: $T_0 - \upsilon L > 0$. That is, the available time is more than sufficient for a resident to travel from the CBD to the boundary of the isolated state.

Maximizing U subject to the budget constraint, (4.1.13), yields

$$c = \xi\Omega, \quad RL_h = \eta\Omega, \quad S = \lambda\Omega - (1 - \delta_k)k, \quad T_{hj} = \frac{\sigma_j\Omega}{w_j},$$
$$j = 1, 2, \quad 0 \leq \omega \leq L$$
(4.1.14)

where

$$\Omega(\omega,t) \equiv \rho(rk + T_0 w + \bar{r}) + \rho_0 k - \upsilon\rho\omega w, \quad w(t) \equiv w_1 + w_2,$$
$$\rho \equiv \frac{1}{\sigma_1 + \sigma_2 + \xi + \eta + \lambda}, \quad \rho_0 \equiv \rho(1 - \delta_k).$$
(4.1.15)

The capital accumulation is given by

$$\frac{dk(\omega)}{dt} = S(\omega) - \delta_k k(\omega), \quad 0 \leq \omega \leq L.$$

Substituting S in (4.1.14) into the above equation yields

$$\frac{dk(\omega)}{dt} = \lambda\Omega(\omega) - \delta_k k(\omega), \quad 0 \leq \omega \leq L.$$
(4.1.16)

The above equation gives the dynamics of capital accumulation of a typical family at location ω.

As the economy is isolated, the total population is distributed over the whole urban area. The population constraint is given by

$$\int_0^L n(\omega,t)\,d\omega = N.$$
(4.1.17)

Similarly, the consumption constraint is given by

$$\int_0^L c(\omega,t)n(\omega,t)\,d\omega = C \qquad (4.1.18)$$

where C is the total consumption of the state. As the industrial product is assumed to be either consumed or invested, we have the following equation

$$\overline{S}(t) + C(t) = F(t) \qquad (4.1.19)$$

where $\overline{S}(t)$ is the total investment (= the total savings) of the society, defined as follows

$$\overline{S}(t) \equiv \int_0^L S(\omega,t)n(\omega,t)\,d\omega . \qquad (4.1.20)$$

As the total capital stocks are equal to the sum of the capital stocks owned by the population, we obtain

$$\int_0^L k(\omega,t)n(\omega,t)\,d\omega = K . \qquad (4.1.21)$$

We have thus built a dynamic growth model with endogenous spatial distribution of wealth, consumption and population, capital accumulation, the sexual division of labor, and residential location. The system has 13 space-time-dependent variables, k, c, L_h, S, n, T_1, T_2, T_{h1}, T_{h2}, A, U, R, and y, and 8 time-dependent variables, F, K, C, w_1, w_2, r, $\overline{S}$, and $\bar{r}$. The system contains 21 independent equations.

4.2 The Spatial Equilibrium Structure

Before discussing whether the dynamic system has a long-run equilibrium, we examine the relations between the husband and wife's wage rates and the time distribution at any point of time.

By (4.1.4), the ratio of the wage rates per unit of time between the husband and wife is given by: $w_1/w_2 = z$. The ratio is equal to their productivity difference z and is independent of the capital stock and production scale. This implies that there is no sexual discrimination in the labor market.

By $w_1/w_2 = z$ and (4.1.14), the ratio of leisure time of the husband and wife is given by

$$\frac{T_{h1}}{T_{h2}} = \frac{\sigma_1}{\sigma_2 z}. \qquad (4.2.1)$$

The ratio is dependent on sexual differences in marginal values of leisure time and productivity. In the case of $\sigma_1 = \sigma_2$, if the husband is more productive than the wife in work outside the home, i.e., $z > 1$, then the husband works longer than the wife. If the wife's family marginal utility of leisure time is higher than the husband's, she tends to stay at home longer than the husband. If there is no sexual difference in the marginal utility of leisure and productivity, then the husband and the wife have an identical time distribution.

The conditions for the existence of equilibria are given as follows.

Proposition 4.2.1.
The dynamic system always has equilibria. The number of equilibria is the same as the number of solutions of the equation, $\Phi(\Lambda) = 0$, where Φ is a function of Λ $(= wT_0 + \eta K/\lambda N)$ defined in Appendix A.4.1.

The above proposition is proved in the appendix of this chapter. As shown in the appendix, the equation, $\Phi(\Lambda) = 0$, has at least one meaningful solution. Moreover, for any solution Λ of $\Phi(\Lambda) = 0$, the variables in the system are uniquely determined as functions of Λ by the following equations

$$K = \frac{\Lambda - T_0 w}{\eta} \lambda N, \qquad (4.2.2)$$

$$C = \frac{\xi L}{\lambda}, \quad \bar{S} = \delta_k K, \qquad (4.2.3)$$

$$F = \xi_0 K, \quad N^* = \xi_0^{1/\beta} K, \qquad (4.2.4)$$

$$r = \alpha \xi_0, \quad w_1 = \beta z \xi_0^{-\alpha/\beta}, \quad w_2 = \beta \xi_0^{-\alpha/\beta}, \qquad (4.2.5)$$

$$\bar{R} = \frac{\eta K}{\lambda}, \quad \bar{r} = \frac{\eta K}{\lambda N}, \qquad (4.2.6)$$

$$n(\omega) = \frac{(\Lambda - v\omega w)^\theta}{\Lambda^\theta} n(0), \quad 0 < \omega \leq L,$$

$$n(0) = \frac{1+\theta}{\Lambda^{\theta+1} - (\Lambda - \upsilon Lw)^{\theta+1}} \upsilon w N \Lambda^{\theta}, \quad (4.2.7)$$

$$k(\omega) = \frac{\Lambda - \upsilon \omega w}{\Lambda} k(0), \quad 0 < \omega \le L,$$

$$k(0) = \frac{\Lambda}{(\sigma_1 + \sigma_2 + \eta)/\lambda + w_2 \xi_0^{1/\beta}}, \quad (4.2.8)$$

$$T_{hj}(\omega) = \frac{\sigma_j k(\omega)}{\lambda w_j}, \quad T_j(\omega) = T_0 - \upsilon \omega - \sigma_j k(\omega)/\lambda w_j, \quad (4.2.9)$$

$$N_j = n_j K, \quad (4.2.10)$$

$$c(\omega) = \frac{\xi k(\omega)}{\lambda}, \quad R(\omega) = \frac{\eta k(\omega) n(\omega)}{\lambda}, \quad 0 < \omega \le L \quad (4.2.11)$$

where

$$\xi_0 \equiv \delta_0 + \frac{\xi}{\lambda}, \quad n_j \equiv \frac{\sigma_1 + \sigma_2}{\lambda w} + \frac{\xi_0^{1/\beta}}{1+z} - \frac{\sigma_j}{\lambda w_j}.$$

The above equations are given in the appendix.

It should be noted that although the system may have multiple equilibria (i.e., $\Phi(\Lambda) = 0$ has multiple solutions), from the above equations one sees that the rate of interest r and the wage rates w_j of the two sexes per unit of working time are invariant at any equilibrium.

4.3 Sexual Productivity Differences and Economic Structure

By (4.2.5), we have

$$\frac{dr}{dz} = 0, \quad \frac{dw_1}{dt} = \frac{w_1}{z} > 0, \quad \frac{dw_2}{dz} = 0. \quad (4.3.1)$$

As the husband's productivity is improved, his wage rate w_1 is increased, but the rate of interest r, and the wife's wage rate per unit of working time w_2 are not affected. It should be noted that in order to explain how the equilibrium value of a variable is affected by shifts in parameters, it is necessary to explain how all equations in the system are affected. As the system consists of many equations, we omit a detailed economic interpretation of the analytical results.

Taking derivatives of (4.A.1.14) with respect to z yields

$$\frac{d\Phi}{d\Lambda}\frac{d\Lambda}{dz} = \upsilon L w_2 k_0 \Lambda_0 \qquad (4.3.2)$$

in which

$$\Lambda_0 = \frac{\upsilon w L \Lambda^{1+\theta}(2+\theta)(\Lambda - \upsilon Lw)^\theta}{\left[\Lambda^{\theta+1} - (\Lambda - \upsilon Lw)^{\theta+1}\right]\left[\Lambda^{\theta+2} - (\Lambda - \upsilon Lw)^{\theta+2}\right]}$$

$$- \frac{(\Lambda - \upsilon Lw)^\theta}{\Lambda^{\theta+1} - (\Lambda - \upsilon Lw)^{\theta+1}} - \frac{T_0}{\Lambda - T_0 w} \qquad (4.3.3)$$

where $d\Phi/d\Lambda$ is defined by (4.A.1.16). From (4.A.1.16), (4.3.2) and (4.3.3), we conclude that the sign of $d\Lambda/dz$ may be either positive or negative.

By (4.2.2), we have

$$\frac{dK}{dz} = \left(\frac{d\Lambda}{dz} - T_0 w_2\right)\frac{\lambda N}{\eta}. \qquad (4.3.4)$$

If $d\Lambda/dz$ is negative, then an improvement in the male population's productivity reduces the level of total capital stocks. If $d\Lambda/dz$ is positive, then dK/dz may be either positive or negative.

By (4.2.3), (4.2.4) and (4.2.6), we have

$$\frac{dC}{dz} = \frac{\xi}{\lambda}\frac{dK}{dz}, \quad \frac{dF}{dz} = \xi_0 \frac{dK}{dz}, \quad \frac{dN^*}{dz} = \xi_0^{1/\beta}\frac{dK}{dz}, \quad \frac{d\bar{r}}{dz} = \frac{\eta}{\lambda N}\frac{dK}{dz} \qquad (4.3.5)$$

If $dK/dz > (<) 0$, then the levels of total consumption of the commodity C, the output F, the total qualified labor input N^*, and the revenue from land ownership per family $\bar{r}$ are increased (reduced).

The impact on residential density is given by

4.3 Sexual Productivity Differences and Economic Structure

$$\frac{1}{n(0)}\frac{dn(0)}{dz} = \frac{1}{1+z} - \frac{\upsilon w_2 L(\Lambda - \upsilon Lw)^\theta}{\Lambda^{\theta+1} - (\Lambda - \upsilon Lw)^{\theta+1}} +$$

$$\left[\frac{\theta}{\Lambda} - \frac{\Lambda^\theta - (\Lambda - \upsilon Lw)^\theta}{\Lambda^{\theta+1} - (\Lambda - \upsilon Lw)^{\theta+1}}(1+\theta)\right]\frac{d\Lambda}{dz},$$

$$\frac{1}{n(\omega)}\frac{dn(\omega)}{dz} = \frac{1}{n(0)}\frac{dn(0)}{dz} - \theta \upsilon w_2 (\Lambda - \upsilon \omega w)^{\theta-1} +$$

$$\left[(\Lambda - \upsilon \omega w)^{\theta-1} - \Lambda^{\theta-1}\right]\theta \frac{d\Lambda}{dz}, \quad 0 < \omega \le L. \tag{4.3.6}$$

From (4.2.8), we get the effects of changes in z on $k(\omega)$ as follows

$$\frac{dk(0)}{dz} = \frac{k(0)}{\Lambda}\frac{d\Lambda}{dz},$$

$$\frac{dk(\omega)}{dz} = \left(\frac{d\Lambda}{dz} - \upsilon \omega w_2\right)\frac{k(\omega)}{\Lambda - \upsilon \omega w}, \quad 0 < \omega \le L. \tag{4.3.7}$$

The effects on the working time of the family at location ω and each sex's total working time are given by

$$\frac{\lambda w_1}{\sigma_1}\frac{dT_1(\omega)}{dz} = \frac{k(\omega)}{z} - \frac{dk(\omega)}{dz}, \quad \frac{\lambda w_2}{\sigma_2}\frac{dT_2(\omega)}{dz} = -\frac{dk(\omega)}{dz},$$

$$\frac{1}{K}\frac{dN_1}{dz} = \frac{\sigma_1}{\lambda z w_1} - \frac{\sigma_1 + \sigma_2}{(1+z)\lambda w} - \frac{\xi_0^{1/\beta}}{(1+z)^2} + \frac{n_1}{K}\frac{dK}{dz},$$

$$\frac{1}{K}\frac{dN_2}{dz} = -\frac{\sigma_1 + \sigma_2}{\lambda w_2} - \xi_0^{1/\beta} + \frac{n_2}{K}\frac{dK}{dz}. \tag{4.3.8}$$

The effects on the family's consumption level and land rent at location ω are given by

$$\frac{dc(\omega)}{dz} = \frac{\eta \xi}{\lambda}\frac{dk(\omega)}{dz},$$

$$\frac{1}{R(\omega)}\frac{dR(\omega)}{dz} = \frac{1}{n(0)}\frac{dn(0)}{dz} - \upsilon w_2(1+\theta)(\Lambda - \upsilon \omega w)^\theta +$$

$$\left[(1+\theta)(\Lambda - \upsilon \omega w)^\theta - \theta \Lambda^{1-\theta}\right]\frac{d\Lambda}{dz}, \quad 0 < \omega \le L. \tag{4.3.9}$$

72   4   Urban Structure with Growth and Sexual Division of Labor

Although the effects of changes in shifts of z are explicitly provided, unfortunately it is not easy to explicitly interpret the analytical results (because of the complex representations of the results).

As all the variables are explicitly given, it is possible to carry out comparative static analysis directly with respect to any parameter. For instance, we may directly examine the effects of changes in the family's propensity to save λ, to consume housing η, to consume the commodity ξ, and to use the husband's (wife's) leisure time $\sigma_1(\sigma_2)$ on the long-run behavior of the system.

4.4   Remarks

This chapter proposed a simple growth model on the basis of the neoclassical one-sector growth model, the Alonso model and family economics. It is well known that each approach has recently been refined and extended in various ways. It is thus conceptually easy to extend the model in this chapter on the basis of the literature on urban economics (e.g., Muth, 1973, Arnott, 1987, Anas, 1982, Brueckner, 1981), neoclassical growth theory, and family economics.

Appendix

A.4.1   Proving Proposition 4.2.1

We now prove Proposition 4.2.1. From (4.1.16), equilibrium is determined as a solution of the following equation

$$\Omega(\omega) = \frac{k(\omega)}{\lambda}, \quad 0 \leq \omega \leq L. \tag{4.A.1.1}$$

Substituting (4.A.1.1) into (4.1.14) yields

$$c = \frac{\xi k}{\lambda}, \quad RL_h = \frac{\eta k}{\lambda}, \quad S = \delta_k k, \quad T_{hj} = \frac{\sigma_j k}{\lambda w_j}. \tag{4.A.1.2}$$

From $c = \xi k / \lambda$, (4.1.18) and (4.1.21), $C = \xi K / \lambda$ is held. Substituting $S = \delta_k k$ into (4.1.20) yields: $\overline{S} = \delta_k K$. We get (4.2.3). Substituting (4.2.3) into (4.1.19) and using $F = K^\alpha N^{*\beta}$, we have (4.2.4). From (4.1.4) and (4.2.4), we get (4.2.5). By $RL_h = \eta k / \lambda$ and (4.1.5), we have

$$R(\omega) = \frac{\eta k(\omega)n(\omega)}{\lambda}, \quad 0 \leq \omega \leq L. \tag{4.A.1.3}$$

Substituting this equation into (4.1.6), we get (4.2.6). On the other hand, substituting (4.1.9) and (4.1.14) into (4.1.8) and using (4.1.5) and $U(0) = U(\omega)$, we have

$$\left[\frac{n(0)}{n(\omega)}\right]^{\mu+\eta} = \left[\frac{\Omega(0)}{\Omega(\omega)}\right]^{1/\rho-\eta}. \tag{4.A.1.4}$$

Substituting $\Omega = k/\lambda$ into (4.A.1.4) yields

$$n(\omega) = n(0)\left[\frac{k(\omega)}{k(0)}\right]^\theta \tag{4.A.1.5}$$

where $\theta \equiv (1-\eta\rho)/\rho(\mu+\eta) > 0$. From the definition of Ω in (2.15) and $\Omega = k/\lambda$, we get

$$k = \frac{T_0 w - \upsilon\omega w + \bar{r}}{1/\lambda - \rho_0 - \rho r}\rho, \quad 0 \leq \omega \leq L \tag{4.A.1.6}$$

in which w, r and $\bar{r}$ are independent of ω. From (4.A.1.6) and (4.2.6), we have

$$k(\omega) = \frac{\Lambda - \upsilon\omega w}{\Lambda}k(0) \tag{4.A.1.7}$$

in which $\Lambda(K) \equiv T_0 w + \eta K/\lambda N > 0$. Substituting (4.A.1.7) into (4.A.1.5) yields $n(\omega)$ in (4.2.7). Substituting $n(\omega)$ in (4.2.7) into (4.1.17), we solve $n(0)$ as (4.2.7). From (4.2.7), we determine $n(\omega)$ at any location ω, $0 \leq \omega \leq L$, as functions of ω and $\bar{r}$. By (4.A.1.7) and (4.2.7), we have

$$k(\omega)n(\omega) = \frac{(\Lambda - \upsilon\omega w)^{1+\theta}}{\Lambda^{1+\theta}}k(0)n(0). \tag{4.A.1.8}$$

Substituting (4.A.1.8) into (4.1.21) yields (4.2.8). From (4.A.1.7) and (4.2.8), we determine $k(\omega)$ at any location ω, $0 \leq \omega \leq L$, as functions of K, w and $\bar{r}$. The time distribution of the husband and wife is given by (4.2.9).

It is easy to see that if the capital stocks K are determined, then the equilibrium values of all other variables are uniquely determined. We now determine the equilibrium values of K.

We may rewrite $T_{hj}(\omega)$ in (4.2.9) as follows

$$T_{hj}(\omega) = \frac{\Lambda - \upsilon \omega w}{w} - \frac{\sigma_j k(\omega)}{\lambda w_j} - \frac{\eta K}{\lambda w N} \qquad (4.A.1.9)$$

in which $\Lambda = T_0 w + \eta K / \lambda N$ is used. Substituting (4.A.1.7) into (4.A.1.1.9) yields

$$T_{hj}(\omega) = \left[\frac{\Lambda}{wk(0)} - \frac{\sigma_j}{\lambda w_j}\right] k(\omega) - \frac{\eta K}{\lambda w N}. \qquad (4.A.1.10)$$

Multiplying (4.A.1.10) by $n(\omega)$ and then integrating the equation from 0 to L, we get

$$N_j = \left[\frac{\Lambda}{wk(0)} - \frac{\sigma_j}{\lambda w_j} - \frac{\eta}{\lambda w}\right] K. \qquad (4.A.1.11)$$

Using (4.A.1.11) and $N^* = zN_1 + N_2$, we obtain

$$\frac{N^*}{1+z} = \left[\frac{\Lambda}{k(0)} - \frac{\sigma_1}{\lambda} - \frac{\sigma_2}{\lambda} - \frac{\eta}{\lambda}\right] \frac{K}{w} \qquad (4.A.1.12)$$

where $w_1 / w_2 = z$ is used. Substituting N^* in (4.A.1.12) into $N^* = \xi_0^{1/\beta} K$ in (4.2.4) yields

$$\frac{\Lambda}{k(0)} = \frac{\sigma_1 + \sigma_2 + \eta}{\lambda} + \frac{w \xi_0^{1/\beta}}{1+z}. \qquad (4.A.1.13)$$

Substituting (4.A.1.13) into (4.A.1.11) yields (4.2.10). The right-hand side of (4.A.1.13) is constant and the left-hand side is a function of K. Substituting (4.2.2), (4.2.7) and (4.2.8) into (4.A.1.13), we get the following equation to determine Λ

$$\Phi(\Lambda) \equiv \frac{\Lambda^{\theta+2} - (\Lambda - \upsilon L w)^{\theta+2}}{(\Lambda - T_0 w)\left[\Lambda^{\theta+1} - (\Lambda - \upsilon L w)^{\theta+1}\right]} - k_0 = 0,$$

$$T_0 w < \Lambda < \infty \quad (4.A.1.14)$$

in which

$$k_0 \equiv \frac{\sigma_1 + \sigma_2 + \eta + \lambda w \xi_0^{1/\beta}/(1+z)}{(1+\theta)\eta}(2+\theta) > 0. \quad (4.A.1.15)$$

We have: $\Phi(T_0 w) > 0$ and

$$\Phi(\infty) = \frac{\theta + 2}{\theta + 1} - k_0.$$

As $(\theta + 2)/(\theta + 1) < k_0$, we have: $\Phi(\infty) < 0$. Accordingly, the equation, $\Phi(\Lambda) = 0$, $T_0 w < \Lambda < \infty$, has at least one solution.

Taking derivatives of (4.A.1.14) with respect to Λ yields

$$\frac{d\Phi}{d\Lambda} = (\Phi_1 + \Phi_2)\Phi_0 \quad (4.A.1.16)$$

where

$$\Phi_0 \equiv \frac{1}{(\Lambda - T_0 w)^2 \left[\Lambda^{\theta+1} - (\Lambda - \upsilon L w)^{\theta+1}\right]} > 0,$$

$$\Phi_1 \equiv \frac{(\Lambda - \upsilon L w)^{\theta}(\upsilon L w)^2 \Lambda^{\theta}}{\Lambda^{\theta+1} - (\Lambda - \upsilon L w)^{\theta+1}}(1+\theta)(\Lambda - T_0 w) > 0,$$

$$\Phi_2 \equiv w(T_0 - \upsilon L)(\Lambda - \upsilon L w)^{\theta+1} - \Lambda^{\theta+1} T_0 w < 0. \quad (4.A.1.17)$$

It is not easy to explicitly judge the sign of $d\Phi/d\Lambda$ for $T_0 w < \Lambda < \infty$. As Φ_1 is positive and Φ_2 is negative, $d\Phi/d\Lambda$ may be either positive or negative. This implies that $\Phi(\Lambda) = 0$ may have multiple solutions.

The proof of Proposition 4.2.1 is thus completed.

5 Dynamic Urban Pattern Formation with Heterogeneous Population

The necessity of building a general dynamic framework of economic geography has been emphasized for a long time. But only a few dynamic models of economic geography have been developed on the basis of economic principles. The following argument by Wilson (1990) still reflects the current situation of regional science: *"A major current concern is with patterns of regional development in a variety of contexts and at a variety of spatial scales. Regional science should be able to provide a theoretical basis for this field but it can be argued that it has failed to do so."* Many economic theories which explain various aspects of the complexity of economic geography have been proposed. For instance, economic growth theories emphasize economic dynamics with capital, population and knowledge accumulation, but neglect spatial characteristics of economic activities (e.g., Burmeister and Dobell, 1970). Some urban economic models take account of endogenous capital and population growth, but neglect spatial structure (e.g., Henderson, 1985, Miyao, 1987a, Richardson, 1972); some others deal with residential structures but neglect economic production and endogenous incomes. It is reasonable to ask whether it is possible to develop a compact framework within which main ideas in these economic theories or schools can be discussed. This chapter tries to propose a dynamic economic growth model with endogenous residential pattern formation, synthesizing the ideas in the standard one-sector neoclassical growth model, the Kaldor-Pasinetti two-group model, the multi-group urban models, the Alonso location model and the Muth housing model.

The previous chapters integrated the one-sector neoclassical growth model and the Alonso location model. However, we assumed that the population is homogenous. The purpose of this chapter is to deal with urban pattern formation with heterogenous population, based on the Kaldor-Pasinetti model. The significance of the Kaldor-Pasinetti two-class model lies in that it explicitly takes account of endogenous wealth and income distribution among various social and economic groups (see, for instance, Sato, 1966, Pasinetti, 1974, Salvadori, 1991, Panico and Salvadori, 1993). But this approach completely neglects location issues. Although the urban economist has proposed multi-group models (e.g., Rose-Ackerman, 1975, 1977, Schnare, 1976, Yellin, 1974, Yinger, 1976, Kern, 1981, Zhang, 1989a, 1993e), these models are basically developed within static frameworks with fixed incomes. These models have not addressed dynamics of wealth and income distribution among various groups and their relative residential location. Our model is developed on the basis of these models.

Housing is a heterogeneous and highly differentiated product. Housing is the largest single component of the expenditure of most households. A dwelling is produced by combination of different commodities and land, involves a variety of services and possesses a variety of characteristics. Housing units with the same market rent may contain quite different combinations of housing attributes. Households value housing units because of the utility that the housing characteristics offer. In urban economics it is commonly assumed that households maximize a utility function defined on two composite commodities, 'housing' and 'all other goods', subject to budgets. Housing is defined in terms of housing services by Muth (1961). The location of dwelling in relation to the potential consumer's workplace is examined by Muth. The model by Herbert and Stevens (1960) analyzes the trade-off between housing and journey to work costs whilst accounting for the effects of some demographic differences in households and the spatial variation of dwelling stocks. They provided the equilibrium optimal solutions to the housing allocation which are obtained by maximization of bid-rents. Senior and Wilson (1974) showed how perhaps more realistic suboptimal solutions by using total observed rents rather than a model derived the maximization.

People's preferences and human capital are different. Heterogeneity in preferences among individuals is an important determinant of spatial structure. Individuals may differentiate quite strongly among particular locations or areas based on differences in local climate or amenities available. These differences may have impact on the location of consumers and producers. They may also create spatial agglomeration economies in consumption and production. There are many studies that incorporate heterogenity into analytical models of spatial structures. The remainder of this chapter is organized as follows. Section 5.1 defines the basic two-group model of capital accumulation and spatial location. Section 5.2 analyzes location of the two groups at any point of time. Section 5.3 provides conditions for existence of equilibria and stability of the dynamic system. Sections 5.4 and 5.5, respectively, examine effects of changes in group 1's savings rate and group 2's human capital, on long-run growth, wealth and income distribution, and location structure of the economic system. Section 5.6 concludes the study. The proofs of the conclusions in Section 5.3 are given in the Appendix. This chapter is based on Zhang (1994e).

5.1 The Urban Growth with Two Groups

We consider an isolated economic system. Here, by isolation we mean that there is neither migration, nor trade, nor flows of ideas between the system and the outside world. The system consists of two, industrial and housing, sectors. The industrial production is similar to that in the one-sector growth neoclassical model (e.g., Burmeister and Dobell, 1970). We assume that the industrial product can be either invested or consumed. The housing production is similar to that in the Muth model (e.g., Muth, 1961). Housing is supplied with combination of capital and land. The system geographically consists of a finite strip of land extending from a fixed central business district (CBD) with constant unit width. The system consists of the CBD and the residential area. We assume that all economic activities are concentrated at the CBD. The residential area is occupied by households. The economic geography is similar to the Alonso model in the sense that the spatial pattern under consideration is described by residential location over space. For simplicity of expression, we assume that the CBD is located at the left-side end of the linear territory, as illustrated in Fig. 2.1.1.

Similarly to the Kaldor-Pasinetti model (e.g., Pasinetti, 1974, Sato, 1966), we classify the population into two groups and neglect any possibility of group transformation. Different from the literature of multi-group models in urban economics, we assume that there is no prejudice in labor and housing markets. Here, by justice in labor markets we mean that any laborer is paid according to "qualified labor", irrespective of to which group the laborer belongs. We also omit group prejudice in residential location. All the markets are characterized by perfect competition. For simplicity, we assume that the people from the same group have the same level of human capital, though people of different groups have different levels of human capital.

The two groups are indexed by, 1 and 2, respectively. We introduce

N_j — the fixed population of group j;

N_j^* — the qualified labor force (to be defined) of group j;

$K_j(t)$ — the capital stocks owned by group j, $j = 1, 2$, at time t; and

$F(t)$ — output of the industrial sector.

The total qualified labor force N^* and the total capital stock $K(t)$ are given by

$$N^* = N_1^* + N_2^*, \quad K = K_1 + K_2. \tag{5.1.1}$$

We assume that labor and capital are always fully employed.

In order to describe the relationship between N_j and N_j^*, we introduce human capital difference index z to distinguish difference in productivity of the two groups. With group 1's human capital as basis of measurement, we assume the relationship as follows

$$N_1^* = N_1, \quad N_2^* = zN_2. \tag{5.1.2}$$

When $z > 1$, we say that group 2 accumulates more human capital than group 1 (in terms of productivity). Here, we neglect possible impact of education, training and other costly learning efforts on N_j^*. Further possible extensions of endogenous human capital are referred to, e.g., Becker (1975) and Zhang (1999a, 2000b).

We select the industrial commodity to serve as numeraire and introduce

L — the fixed length of the system;
ω — the distance from the CBD to a point in the residential area;
$R(\omega,t)$ — the land rent at location ω;
$w_j(t)$ and $r(t)$ — the wage rate of group j ($j = 1, 2$) and the interest rate, respectively;
$n_j(\omega,t)$ — the residential density of group j ($j = 1, 2$) at location ω;
$c_j(t)$ — the level of commodity consumption per household of group j at location ω, $j = 1, 2$;
$C_j(t)$ — the total consumption level of group j, $j = 1, 2$,; and
$L_h(\omega,t)$ — the lot size per household at location ω.

We now define the basic model.

Production and capital accumulation
We specify the production function as follows

$$F = K_i^\alpha N^{*\beta}, \quad \alpha + \beta = 1, \quad \alpha, \beta > 0 \tag{5.1.3}$$

where $K_i(t)$ is capital stock employed by the industrial sector. The marginal conditions are given by

$$r = \frac{\alpha F}{K_i}, \quad w_1 = \frac{\beta F}{N^*}, \quad w_2 = \frac{z\beta F}{N^*}. \tag{5.1.4}$$

From (5.1.4), we directly have the following result.

Lemma 5.1.1.
The ratio w_1 / w_2 of the wage rates between groups 1 and 2 is equal to the ratio of human capital $1/z$ between the two groups. In particular, when the two groups accumulate the same level of human capital, i.e., $z = 1$, their wage rates are equal.

Lemma 5.1.1 states that the difference in wage rates between the two groups is due to the difference in their human capital. As

$$w = w_1 = \frac{w_2}{z} \tag{5.1.5}$$

we may interpret w as the wage rate per unity of qualified labor force. This equation implies that there is no group prejudice in the labor market.

Land ownership is an extremely complicated factor in modeling. For simplicity of analysis, we assume that land is owned by the state. Land is rented out to residents by auctioning. The state spends the rent income on supplying purely public goods such as military defense and health care systems which have no direct impact upon consumers' and producers' behavior. Under this assumption, the rent income does not enter into households' individual income. This assumption is similar to that of absentee landlords (e.g., Mills and Hamilton, 1985, Fujita, 1989). Surely, we admit that we accept this assumption mainly for convenience of analysis.

The income $Y_j(t)$ of group j, $j = 1, 2$, consists of the interest payment of wealth and the wage rate, i.e.

$$Y_j = rK_j + w_j N_j. \tag{5.1.6}$$

Capital accumulation is thus given by

$$\frac{dK_j}{dt} = s_j Y_j - \delta_k K_j, \quad j = 1, 2 \tag{5.1.7}$$

in which δ_k is the fixed depreciation rate of capital and s_j is the savings rate of group j.

Housing services production
Similar to Chapter 2, we assume that the housing industry supplies housing services by combining land and capital. Denote $c_h(\omega,t)$ housing service received by the household at location ω. We specify the housing service production functions as follows

$$c_h(\omega,t) = K_h^{\alpha_h} L_h^{\beta_h}, \quad \alpha_h + \beta_h = 1, \quad \alpha_h, \beta_h \tag{5.1.8}$$

where $L_h(\omega,t)$ and $K_h(\omega,t)$ are input levels of land and capital per household at location ω, respectively. The marginal conditions are given by

$$r = \frac{\alpha_h R_h c_h}{K_h}, \quad R = \frac{\beta_h R_h c_h}{L_h}, \quad 0 \le \omega \le L \tag{5.1.9}$$

where $R_h(\omega,t)$ is the housing rent at location ω at time t.

Consumption and location choice of the households
We assume that the level of utility that a household can obtain from consuming commodity and housing can be expressed in the following form

$$U_j(\omega,t) = c_j^{\xi_j}(\omega,t) c_{hj}^{\eta_j}(\omega,t), \quad \xi_j, \eta_j > 0, \quad j = 1, 2 \tag{5.1.10}$$

where $c_{hj}(\omega,t)$ is consumption level of housing services per household of group j at location ω at time t. The consumption budget of group j is given by $(1 - s_j) Y_j / N_j$. The consumer problem is defined by

$$\max U_j(\omega,t), \quad \text{s.t:} \quad c_j(\omega) + R_h(\omega) c_{hj}(\omega) = \frac{(1 - s_j) Y_j}{N_j} - \Gamma(\omega) \tag{5.1.11}$$

where $\Gamma(\omega)$ is the total traveling cost between dwelling site ω and the CBD.

The balance of demand for and supply of the industrial product is given by

$$C_1 + s_1 Y_1 + C_2 + s_2 Y_2 = F. \tag{5.1.12}$$

We assume that the resident traveling from ω to the CBD is charged in the following way

$$\Gamma(\omega) = \upsilon \omega, \quad \upsilon > 0, \quad 0 \le \omega \le L \tag{5.1.13}$$

where υ is the constant travel cost per unity of distance. The problem has the following unique solution

$$c_j(\omega) = \xi_j \bar{s}_j y_j(\omega), \quad c_{hj}(\omega) = \frac{\eta_j \bar{s}_j y_j(\omega)}{R_h(\omega)}, \quad j = 1, 2 \tag{5.1.14}$$

in which

$$\overline{s}_j \equiv \frac{1-s_j}{\xi_j+\eta_j}, \quad y_j(\omega,t) \equiv \overline{S}_j Y_j(t) - \Gamma(\omega), \quad \overline{S}_j \equiv \frac{1-s_j}{N_j}.$$

We call $y_j(\omega,t)$ the net consumption level of the household of group j at location ω and at time t. In the remainder of this chapter, we denote $y_j(t) = y_j(0,t)$. That is, $y_j(t)$ is the net consumption level of the household of group j at the boundary between the residential area and the CBD at time t. We require

$$y_j(L,t) = y_j(t) - \upsilon L > 0.$$

That is, the total consumption budget of any household is always sufficient for traveling from the CBD to the boundary of the isolated state.

According to the definitions, we have

$$n_j(\omega,t) = \frac{1}{L_h(\omega,t)}, \quad 0 \leq \omega \leq L, \quad j = 1, 2. \tag{5.1.15}$$

As the state is isolated and the population is constant, the total population is distributed over the whole urban area. The population constraints are given by

$$\int_0^L n_j(\omega,t)d\omega = N_j, \quad j = 1, 2. \tag{5.1.16}$$

Similarly, the consumption constraints are given by

$$c_h(\omega,t) = c_{hj}(\omega,t), \quad 0 \leq \omega \leq L, \quad j = 1, 2,$$

$$\int_0^L c_j(\omega,t)n_j(\omega,t)d\omega = C_j(t), \quad j = 1, 2. \tag{5.1.17}$$

Full employment of capital
The assumption that capital is always fully employed is represented by

$$K_i(t) + \int_0^L \sum_j K_h(\omega,t) n_j(\omega,t) d\omega = K(t). \tag{5.1.18}$$

We have thus completed the dynamic two-group model with capital accumulation, income and wealth distribution and residential location. The system consists of 27 endogenous time-dependent variables, K_j, $n_j(\omega)$, $c_j(\omega)$, $c_h(\omega), c_{hj}(\omega)$, C_j, w_j, Y_j, $U_j(\omega)$ $(j = 1, 2)$, K_i, $K_h(\omega)$, $L_h(\omega)$, K, F, R, $R_h(\omega)$, r, and w. It also contains the same number of independent equations.

5.2 Separation of the Groups' Residential Location

It is important to examine whether the two groups co-exist in the same location. If the two groups are separately located, we have to decide which group is located closer to the CBD. This section solves this problem.

As the households of the same group have identical utility function, the utility level U_j is identical over space within each group. That is

$$U_j(\omega_1) = U_j(\omega_2), \quad 0 \leq \omega_1, \omega_2 \leq L, \quad j = 1, 2.$$

Substituting (5.1.14) into (5.1.10) and using $U_j(\omega_1) = U_j(\omega_2)$, we get

$$\frac{R_h(\omega_1)}{R_h(\omega_2)} = \left[\frac{y_j(\omega_1)}{y_j(\omega_2)}\right]^{n_j}, \quad j = 1, 2 \tag{5.2.1}$$

in which $n_j \equiv 1 + \xi_j / \eta_j$. This equality holds for any ω_1 and ω_2, $0 \leq \omega_1, \omega_2 \leq L$, $j = 1, 2$, at any point of time. From (5.2.1), we see that if $\overline{S}_1 Y_1 = \overline{S}_2 Y_2$, the two groups may co-exist. But if the consumption levels per household $\overline{S}_j Y_j$ are not equal between the two groups, the equality

$$\frac{y_1(\omega_1)}{y_1(\omega_2)} = \frac{y_2(\omega_1)}{y_2(\omega_2)}, \quad y_j(\omega_k) = y_j - \upsilon \omega_k \tag{5.2.2}$$

cannot hold. This implies that it is impossible for the two groups to co-exist at any two points in the whole urban area. This means that there is a distance $L^*(t)$,

5 Dynamic Urban Pattern Formation with Heterogeneous Population

$0 < L^*(t) < L$, such that the two groups are separately located within $[0, L^*)$ and $[L^*, L]$. We now define a rule for determining which of the two groups is located closer to the CBD.

Utilizing (5.1.10) and (5.1.14), we have

$$R_h(L^*, t) = \eta_j \bar{s}_j (\xi_j \bar{s}_j)^{\xi_j/\eta_j} U_j^{*-1/\eta_j} (y_j - \upsilon L^*)^{\eta_j} \qquad (5.2.3)$$

where U_j^* is the utility level evaluated at $\omega = L^*$. In order to determine which group is closer to the CBD, we introduce the following rule well known in the literature of urban economics (e.g., Fujita, 1989).

The rule of groups' relative location:
If housing rent function of group k is steeper than that of group j at the intersection, then the equilibrium location of group k is closer to the CBD than that of group j.

Taking partial derivatives of (5.2.1) with respect to L^* yields

$$\frac{\partial R_h(L^*, U_j^*)}{\partial L^*} = -\frac{n_j R_h(L^*)}{y_j(L^*)}. \qquad (5.2.4)$$

Define

$$\Delta \equiv \frac{\partial R_h(L^*, U_1^*)/\partial L^* - \partial R_h(L^*, U_2^*)/\partial L^*}{R_h(L^*)} = \frac{n_2}{y_2(L^*)} - \frac{n_1}{y_1(L^*)}. \qquad (5.2.5)$$

According to the rule, we see that if $\Delta > 0 (< 0)$, group 2 (group 1) is located near to the CBD. Fig. 5.1.1 illustrates the case of $\Delta > 0$.

Summarizing the above discussions, we have the following result.

Lemma 5.2.1.
If $y_1(L^*)/y_2(L^*) \neq n_1/n_2$, then the two groups are separately located in the urban area. If $y_1(L^*)/y_2(L^*) > n_1/n_2$, group 2 is located closer to the CBD, and vice versa.

5.2 Separation of the Groups' Residential Location

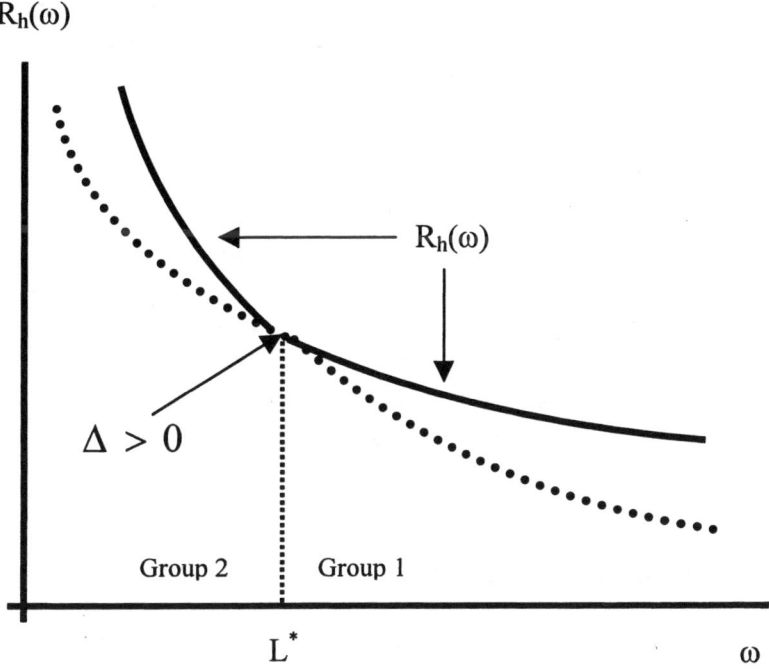

Fig. 5.2.1. The Location of the Two Groups

In the remainder of this chapter, we assume $y_1(L^*)/y_2(L^*) \neq n_1/n_2$ at any point of time. The conclusion about the separation of residents with various incomes is well known from the literature of urban economics. To interpret the conditions for relative location, we notice: $y_j(L^*) = (1 - s_j)Y_j/N_j - \upsilon L^*$ and $n_j \equiv 1 + \xi_j/\eta_j$. We see that the relative location is determined by differences in incomes per household, savings rates, and preferences between the two groups. Although it is not easy to explicitly interpret the conditions, let us examine some special cases.

In the case of $s_1 = s_2$ and $n_1 = n_2$, i.e., the savings rates and preference structures of the two groups being identical, the sign of Δ is identical to that of $Y_1/N_1 - Y_2/N_2$. In this case, if group 2 is richer than group 1 in terms of per household, then group 2 will be further located away from the CBD than group 1. Another case is $n_1 = n_2$ and $Y_1/N_1 = Y_2/N_2$. It is easy to check that if $s_1 < s_2$, group 2 is located closer to the CBD than group 1. That is, the preferences and incomes per household are identical between the two groups, the group with higher savings rate is located closer to the CBD. Similarly, in the case of

$(1-s_1)Y_1/N_1 = (1-s_2)Y_2/N_2$, if $n_1 < n_2$ (i.e., $\xi_2/\eta_2 > \xi_1/\eta_1$), then group 2 is located closer to the CBD than group 1. For convenience of discussion, we make the following assumption.

Assumption 5.2.1.
In the remainder of this study, we assume that the two groups have an identical preference structure (i.e., $n \equiv n_1 = n_2$) and the inequality, $y_1(t,L^*)/y_2(t,L^*) < n_1/n_2 = 1$, is held at any point of time.

Under this assumption, group 1 is always located closer to the CBD. It should be remarked that this requirement is only for convenience of discussion. It can be seen that when the two groups have an identical saving rate (i.e., $s_1 = s_2$), and group 2's human capital and initial capital are higher than group 1's, then the requirement in Assumption 5.2.1 is satisfied.

5.3 Economic Equilibrium and Stability

This section examines whether the system has equilibria and under what conditions it is stable. We find equilibrium by

$$s_j Y_j = \delta_k K_j, \quad j = 1, 2. \tag{5.3.1}$$

According to the definitions of Y_j, we have

$$Y_1 = \frac{\alpha K_1 F}{K_i} + \frac{\beta N_1 F}{N^*}, \quad Y_2 = \frac{\alpha K_2 F}{K_i} + \frac{\beta z N_2 F}{N^*} \tag{5.3.2}$$

in which we use (5.1.4) and F is a function of K_i. To solve (5.3.1), it is sufficient to express K_i as functions of K_1 and K_2. From (5.3.1) and (5.3.2), we get

$$\frac{\delta K_1}{s_1} = \frac{\alpha K_1 F}{K_i} + \frac{\beta N_1 F}{N^*}, \quad \frac{\delta K_2}{s_2} = \frac{\alpha K_2 F}{K_i} + \frac{\beta z N_2 F}{N^*}. \tag{5.3.3}$$

In the appendix, we prove that K_i is determined by the following equation

$$\Phi(K^*) \equiv (\delta u_0 K^* + 1)N_1 S_1 + z N_2 S_2 - \frac{1+u_0}{\beta} N^* = 0 \tag{5.3.4}$$

5.3 Economic Equilibrium and Stability

in which

$$S_j(L^*) \equiv \frac{1}{\delta K^*/s_j - \alpha} > 0, \quad u_0 \equiv \frac{\beta_n \eta_2}{\alpha \xi_2}, \quad K^* \equiv \frac{K_i}{F} = \left(\frac{K_i}{N^*}\right)^\beta.$$

In (5.3.4), there is only one variable K^*. There are three independent equations in (5.3.3) and (5.3.4) and three variables K_1, K_2, and K_i. We now show that from these three equations we can uniquely determine the three variables. The following lemma is proved in Appendix 5.A.1.

Lemma 5.3.1.
The equations, (5.3.3) and (5.3.4), uniquely determine the equilibrium values of K_1, K_2 and K_i.

From (5.1.3), (5.1.4) and (5.1.6), we directly have F, r, w_j, and Y_j ($j = 1, 2$) as functions of the parameters in the system. We now determine the other variables in the system. As shown in the appendix, we have first to determine the separation point L^* of the two groups. Substituting (5.A.1.8) into (5.A.1.10), we obtain the following equation

$$G(L^*) \equiv \frac{1 - Z_1^{\lambda+2}}{1 - Z_1^{\lambda+1}} + \frac{1 - Z_2^{\lambda+2}}{1 - Z_2^{\lambda+1}} y_0 - \xi_0 = 0 \tag{5.3.5}$$

where

$$\lambda \equiv n - \alpha_n, \quad Z_1(L^*) \equiv \frac{y_1(L^*)}{y_1}, \quad Z_2(L^*) \equiv \frac{y_2(L^*)}{y_2(L)},$$

$$y_0 \equiv \frac{\xi_2 \bar{s}_2 y_2(L) N_2}{\xi_1 \bar{s}_1 y_1(L) N_1}, \quad \xi_0 \equiv \frac{(2 + \lambda)(F - \delta K_1 - \delta K_2)}{(1 + \lambda)\xi_1 \bar{s}_1 y_1(L) N_1}. \tag{5.3.6}$$

From the definitions of $Z_1(L^*)$ and $Z_2(L^*)$, we see that it is not easy to explicitly determine the existence of a positive L^*, $0 < L^* < L$, such that $G(L^*) = 0$ is held. Here, we make the following assumption.

Assumption 5.3.1.
In the remainder of this chapter, we assume that the equation, $G(L^*) = 0$, $0 < L^* < L$, has at least one solution.

Although the equation, $G(L^*) = 0$, is explicitly defined, it is not easy to interpret this requirement. We now show that when L^* is determined, all the other location-dependent variables are uniquely determined. As K_j, Y_j, K_i and L^* are given as functions of the parameters in the system, from (5.A.1.6) we directly determine the residential densities, $n_1(0)$ and $n_2(L)$. From this and (5.A.1.5), we solve the residential density, $n_1(\omega)$, $0 < \omega < L^*$, $n_2(\omega)$, $L^* < \omega < L$, at any location. The land use per household at any location ω is directly given by $L_h(\omega) = 1/n_j(\omega)$. From (5.A.1.8), we have the total consumption levels, C_1 and C_2, of the industrial good by the two groups. From (5.A.1.1), we have capital stock, $K_h(\omega)$, used by the household at location ω, $0 < \omega < L$. From (5.1.14), we directly have the consumption level $c_j(\omega)$ of the industrial commodity per household at any location. From $c_j(\omega)$ and $c_h(\omega)$, we directly determine the utility level $U_j(\omega)$ at any location. As we solved $K_h(\omega)$ and $L_h(\omega)$, from (5.1.8) we can thus determine housing level, $c_h(\omega)$ per household at any location. From (5.1.14), we solve the housing rent at any location by $R_h(\omega) = \eta_j \bar{s}_j y_j(\omega) / c_h(\omega)$. The land rent $R(\omega)$ at any location is directly given by $R = \beta_h R_h c_h / L_h$.

Summarizing the above discussion, we see that we have explicitly solved the 27 endogenous variables as functions of the parameters in the system. Although we guaranteed the existence of the unique equilibrium values of the total capital stock, the net income, wage rate of each group, the industrial output and interest rate, it is not easy to prove whether the other location-dependent variables are unique as we could not prove the existence of a unique separation point L^*. But for any given L^*, all the variables in the system are uniquely determined. This implies that although the net income of each group is uniquely determined, when $G(L^*) = 0$ has multiple solutions there are multiple ways to spend the net incomes, depending upon the savings behavior, preference structures and the transportation conditions.

We now provide the conditions for stability of the dynamic system.

Lemma 5.3.2.
If Δ_2 is positive, where Δ_2 is a parameter defined in (5.A.2.5), the system is stable.

From the definition of Δ_2, it is generally acceptable to assume Δ_2 to be positive. For instance, in the case of $\beta \geq \alpha$, Δ_2 is positive. Even in the case of $\beta < \alpha$, if

$$\left(\frac{K_i}{N^*} - \frac{K_1}{N_1}\right)\left(\frac{K_i}{N^*} - \frac{K_2}{N_2}\right) > 0$$

(i.e., $K_i / N^* < \min \{K_1 / N_1, K_2 / N_2\}$, then Δ_2 is positive. The condition $K_i / N^* < \min \{K_1 / N_1, K_2 / N_2\}$ can be satisfied if the difference of wealth per household between the two groups is not large. We may thus generally conclude that the dynamic system is stable.

5.4 The Impact of Savings Rates

This section examines effects of changes in savings behavior on the system. Here, we are concerned with effects of only one group's savings behavior, e.g., s_1. Although we explicitly give the effects of changes in s_1 on all the variables in the system, we will not interpret the economic implications in detail as it takes a long space to follow how a shift in the parameter affects all the 27 equations and how a new equilibrium is settled after the disturbance.

Taking derivatives of (5.3.4) with respect to s_1 yields

$$-\Phi'\frac{dK^*}{ds_1} = (\delta u_0 K^* + 1)\delta N_1 S_1^2 K^* > 0 \tag{5.4.1}$$

where $\Phi' (< 0)$ is the derivative of Φ with respect to K^* defined in (5.A.1.16). An increase of group 1's propensity to savings increases the total capital of the society. From $K^* = (K_i / N^*)^\beta$, we have

$$\frac{dK_i}{ds_1} = \frac{K_i}{\beta K^*}\frac{dK_i}{ds_1} > 0. \tag{5.4.2}$$

That is, as group 1's savings rate is increased, the total capital employed by the industrial sector is increased. Taking derivatives of (5.A.1.12) with respect to s_1, we have

90   5 Dynamic Urban Pattern Formation with Heterogeneous Population

$$\frac{1}{K_2}\frac{dK_2}{ds_1} = -\frac{(1+\delta/\beta s_2 K^*)\alpha S_2}{\beta K^*}\frac{dK^*}{ds_1} < 0,$$

$$\frac{1}{K_1}\frac{dK_1}{ds_1} = \frac{dK}{ds_1} - \frac{dK_2}{ds_1} > 0. \tag{5.4.3}$$

The capital stock owned by group 2 (1) is reduced (increased). A higher propensity to save of group 1 may not benefit group 2.

Taking derivatives of $F = K_i/K^*$, (5.1.4) and (5.3.1) with respect to s_1, we obtain the effects on F, r, w_j, and Y_j as follows

$$\frac{1}{F}\frac{dF}{ds_1} = \frac{\alpha F}{\beta K^*}\frac{dK^*}{ds_1}, \quad \frac{dr}{ds_1} = -\frac{\alpha}{K^{*2}}\frac{dK^*}{ds_1} < 0,$$

$$\frac{1}{w_j}\frac{dw_j}{ds_1} = \frac{1}{F}\frac{dF}{ds_1} < 0, \quad \frac{1}{Y_1}\frac{dY_1}{ds_1} = \frac{\alpha S_2}{S_1} - \frac{(1+\delta/\beta s_1 K^*)\alpha S_1}{\beta K^*}\frac{dK^*}{ds_1},$$

$$\frac{dY_2}{ds_1} = \frac{\delta}{S_2}\frac{dK_2}{ds_1}. \tag{5.4.4}$$

As group 1's savings rate is increased, the output of the industrial sector and the wage rates of the two groups increase, and the interest rate and the net income of group 2 decline. In the case that $(1/K^*)dK^*/ds_1$ is small, group 1's income may increase.

To determine effects on the location-dependent variables, we examine the impact of changes in s_1 upon the separation point L^*. Taking derivatives of (5.4.1) with respect to s_1, we can explicitly give dL^*/ds_1 as a function of the parameters in the system. But as the expression of dL^*/ds_1 is too complicated, we omit the representation.

Taking derivatives of (5.A.1.6) with respect to s_1 yields

$$\frac{1}{n_1(0)}\frac{dn_1(0)}{ds_1} = \frac{y_1 n_1(0)Z_1^\lambda}{\upsilon N_1}\frac{dZ_1}{ds_1} - \frac{\overline{S}_1}{y_1}\frac{dY_1}{ds_1} + \frac{Y_1}{N_1},$$

$$\frac{1}{n_2(L)}\frac{dn_2(L)}{ds_1} = -\frac{y_2(L)n_2(L)Z_2^{\lambda}}{\upsilon N_2}\frac{dZ_2}{ds_1} - \frac{\overline{S}_2}{y_2(L)}\frac{dY_2}{ds_1} \qquad (5.4.5)$$

where

$$\frac{dZ_1}{ds_1} = \frac{\upsilon L^*}{\overline{S}_1 Y_1^2}\frac{dY_1}{ds_1} - \frac{\upsilon}{\overline{S}_1 Y_1^2}\frac{dL^*}{ds_1} - \frac{\upsilon L^* N_1}{(1-s_1)^2 Y_1},$$

$$\frac{dZ_2}{ds_1} = \frac{\upsilon L^* \overline{S}_2}{y_1^2(L)}\frac{dY_2}{ds_1} - \frac{\upsilon}{y_2(L)}\frac{dL^*}{ds_1}. \qquad (5.4.6)$$

From (5.A.1.5), we have the impact on the residential density as follows

$$\frac{1}{n_1(\omega)}\frac{dn_1(\omega)}{ds_1} = \frac{1}{n_1(0)}\frac{dn_1(0)}{ds_1} + \frac{\lambda\upsilon\omega}{y_1(\omega)}\frac{dY_1}{ds_1} - \frac{\lambda\upsilon\omega y_1 N_1}{(1-s_1)^2 y_1(\omega)Y_1},$$
$$0 < \omega \leq L^*,$$

$$\frac{1}{n_2(\omega)}\frac{dn_2(\omega)}{ds_1} = \frac{1}{n_2(L)}\frac{dn_2(L)}{ds_1} - \frac{\lambda\upsilon\overline{S}_2(L-\omega)}{y_2(\omega)y_2(L)}\frac{dY_2}{ds_1}, L^* < \omega \leq L.$$
(5.4.7)

Taking derivatives of (5.A.1.1) with respect to s_1 yields

$$\frac{1}{K_h(\omega)}\frac{dK_h(\omega)}{ds_1} = \frac{1}{K^*}\frac{dK^*}{ds_1} - \frac{1}{1-s_1} - \frac{Y_1}{N_1 y_1(\omega)} + \frac{\overline{S}_1}{y_1(\omega)}\frac{dY_1}{ds_1},$$
$$0 < \omega \leq L^*,$$

$$\frac{1}{K_h(\omega)}\frac{dK_h(\omega)}{ds_1} = \frac{1}{K^*}\frac{dK^*}{ds_1} + \frac{\overline{S}_2}{y_1(\omega)}\frac{dY_2}{ds_1}, L^* < \omega \leq L. \qquad (5.4.8)$$

From (5.4.6), (5.4.7), and $L_h(\omega) = 1/n_j(\omega)$, we get the effects of changes in group 1's savings rate on land use per household at any location. From $c_h(\omega) = K_h^{\alpha_h}(\omega)/n_j^{\beta_h}(\omega)$, (5.4.7) and (5.4.8), we get $dc_h(\omega)/ds_1$ at any location. From $c_j(\omega) = \xi_j \overline{s}_j y_j(\omega)$, we have the impact upon consumption level per household as follows

$$\frac{1}{c_1(\omega)}\frac{dc_1(\omega)}{ds_1} = \frac{\bar{S}_1}{y_1(\omega)}\frac{dY_1}{ds_1} - \frac{Y_1}{y_1(\omega)N_1} - \frac{1}{1-s_1}, \quad 0 < \omega \le L^*,$$

$$\frac{1}{c_2(\omega)}\frac{dc_2(\omega)}{ds_1} = \frac{\bar{S}_2}{y_2(\omega)}\frac{dY_2}{ds_1}, \quad L^* < \omega \le L. \qquad (5.4.9)$$

Taking derivatives of (5.A.1.8), we directly get dC_j / ds_1. Utilizing the relations

$$R_h(\omega) = \frac{\eta_j \bar{s}_j y_j(\omega)}{c_h(\omega)}, \quad R(\omega) = \frac{\beta_h R_h(\omega) c_h(\omega)}{L_h(\omega)}, \quad 0 \le \omega \le L$$

and the above analytical results, we get the effects of changes in group 1's savings rate on the land rent and housing rent at any residential location.

We have thus obtained the effects of changes in group 1's savings rate on all the endogenous variables in the system. We see that it is very difficult to explicitly interpret the analytical results as we have taken so many interfactional forces in our modeling. The resulted effects are dependent upon relative values of various forces on the system.

5.5 The Impact of Human Capital

This section examines the impact of changes in human capital index z on the long-run economic growth and residential pattern. Taking derivatives of (5.3.4) with respect to z yields

$$-\frac{\beta \Phi'}{N_2 S_2}\frac{dK^*}{dz} = \frac{\beta s_2 u_0 - u_0 - 1}{s_2}\delta K^* + 1 + \alpha u_0. \qquad (5.5.1)$$

As $\beta s_2 u_0 - u_0 - 1$ is negative, we see that dK^* / dz may be either positive or negative. From (5.5.1) and $K^* = (K_i / N^*)^\beta$, we have

$$\frac{1}{K_i}\frac{dK_i}{dz} = \frac{1}{\beta K^*}\frac{dK^*}{dz} + \frac{N_2}{N^*}. \qquad (5.5.2)$$

We assume that an increase in group 2's human capital decreases K^* but increases K_i, i.e., $dK^* / dz < 0$, $dK_i / dz > 0$. This seems to be an acceptable requirement. From (5.A.1.12), we have the impact on the capital stocks owned by the two groups as follows

$$\frac{1}{K_1}\frac{dK_1}{dz} = -\left(\frac{1+\delta/\beta s_1 K^*}{\beta K^*}\right)\alpha s_1 \frac{dK^*}{dz},$$

$$\frac{1}{K_2}\frac{dK_2}{dz} = -\left(\frac{1+\delta K^*/s_2}{\beta K^*}\right)\alpha s_2 \frac{dK^*}{dz} + \frac{1}{z}. \quad (5.5.3)$$

We see that an improvement in group 2's human capital will expand the capital stocks of the two groups. From $F = K_i/K^*$, (5.1.4) and (5.3.1), we obtain the effects on F, r, w_j and Y_j as follows

$$\frac{1}{F}\frac{dF}{dz} = \frac{\alpha}{\beta K^*}\frac{dK^*}{dz} + \frac{N_2}{N^*}, \quad \frac{dr}{dz} = -\frac{\alpha}{K^{*2}}\frac{dK^*}{dz} > 0,$$

$$\frac{1}{w_1}\frac{dw_1}{dz} = \frac{\alpha}{\beta K^*}\frac{dK^*}{dz}, \quad \frac{1}{w_2}\frac{dw_2}{dz} = \frac{\alpha}{\beta K^*}\frac{dK^*}{dz} + \frac{1}{z},$$

$$\frac{1}{Y_j}\frac{dY_j}{dz} = \frac{\delta}{s_j}\frac{dK_j}{dz} > 0. \quad (5.5.4)$$

As group 2's human capital is improved, the incomes of the two groups are increased, even though the impact on group 1's wage rate is negative and the impact on group 2's wage rate may be either negative or positive. The total output of the industrial sector and the interest rate are increased.

Similarly to the previous section, it is not difficult to explicitly determine the impact of changes in z upon all the other variables in the system. We omit the calculation.

5.6 On Urban Evolution with Multiple Groups

This chapter proposed an economic growth model with location choice and land use pattern. The model may be considered as a synthesis of the Solow-Swan one-sector growth model, the Kaldor-Pasinetti two-group model, the Alonso location model and the Muth housing model. The model examined the dynamic interdependence among human capital, transportation conditions, savings behavior, location choice and residential pattern in a just two-group economy. The basic force of economic growth is "neo-classical" in the sense that the dynamics are determined by capital accumulation which is dependent on residents' savings in the long term. We also

94   5 Dynamic Urban Pattern Formation with Heterogeneous Population

examined the impact of changes in group 1's savings rate and group 2's human capital on the variables in the system.

This chapter is the first attempt to develop a compact synthetic economic framework with heterogeneous population in which the main ideas in growth theory and urban economics can be discussed. Irrespective of many obvious strict assumptions, the framework suggested in this chapter is, from a structural point of view, general in the sense that some well-known models in theoretical economics can be considered as special cases of our model. According to the literature in growth theory and urban economics (e.g., Burmeister and Dobell, 1970, Grossman and Helpman, 1991, Henderson, 1985, Zhang, 1992b, 1993c), we see that it is conceptually easy to refine, extend and generalize the model on the basis of the current literature.

Appendix

A.5.1   Proving Lemma 5.3.1.

As interest rate is independent of location, from (5.1.4) and (5.1.9) we have

$$r = \frac{\alpha_h R_h c_h}{K_h} = \frac{\alpha F}{K_i}.$$

We can thus express $K_h(\omega)$ as follows

$$K_h(\omega) = \frac{\alpha_h R_h(\omega) c_h(\omega) K_i}{\alpha F}.$$

Substituting $R_h(\omega)c_h(\omega) = R_h(\omega)c_{hj}(\omega) = \eta_j \bar{s}_j y_j(\omega)$ into this equation yields

$$K_h(\omega) = \frac{\alpha_h \eta_j \bar{s}_j y_j(\omega) K_i}{\alpha F}. \tag{5.A.1.1}$$

From (5.1.9) and $c_{hj}(\omega,t) = K_h^{\alpha_h} L_h^{\beta_h}$, we have

$$\frac{R_h(\omega)}{R_h(0)} = \left[\frac{K_h(\omega)L_h(0)}{K_h(0)L_h(\omega)}\right]^{\alpha_h}, \quad 0 < \omega \leq L^*,$$

$$\frac{R_h(\omega)}{R_h(L)} = \left[\frac{K_h(\omega)L_h(L)}{K_h(L)L_h(\omega)}\right]^{\alpha_h}, \quad L^* < \omega \leq L. \tag{5.A.1.2}$$

Substituting (5.A.1.1) into (5.A.1.2) yields

$$\frac{R_h(\omega)}{R_h(0)} = \left[\frac{y_1(\omega)L_h(0)}{y_1 L_h(\omega)}\right]^{\alpha_h}, \quad 0 < \omega \leq L^*,$$

$$\frac{R_h(\omega)}{R_h(L)} = \left[\frac{y_2(\omega)L_h(L)}{y_2(L)L_h(\omega)}\right]^{\alpha_h}, \quad L^* < \omega \leq L. \tag{5.A.1.3}$$

From (5.2.1) we have

$$\frac{R_h(\omega)}{R_h(0)} = \left[\frac{y_1(\omega)}{y_1}\right]^n, \quad 0 < \omega \leq L^*,$$

$$\frac{R_h(\omega)}{R_h(L)} = \left[\frac{y_2(\omega)}{y_2(L)}\right]^n, \quad L^* < \omega \leq L. \tag{5.A.1.4}$$

The left-hand sides of the equations in (5.A.1.2) and (5.A.1.3) are identical. From (5.1.15), (5.A.1.3) and (5.A.1.4) we thus have the following equations

$$n_1(\omega) = n_1(0)\left[\frac{y_1(\omega)}{y_1}\right]^\lambda, \quad 0 < \omega \leq L^*,$$

$$n_2(\omega) = n_2(L)\left[\frac{y_2(\omega)}{y_2(L)}\right]^\lambda, \quad L^* < \omega \leq L. \tag{5.A.1.4}$$

Substituting (5.3.7) into (5.1.16), we get

$$n_1(0) = \frac{1}{y_1^*(L^*)}, \quad n_2(L) = \frac{1}{y_2^*(L^*)} \tag{5.A.1.6}$$

in which

$$y_1^*(L^*) = \frac{1 - Z_1^{1+\lambda}}{(1 + \lambda)\upsilon N_1} y_1 \geq 0,$$

$$y_2^*(L^*) = \frac{Z_2^{1+\lambda} - 1}{(1 + \lambda)\upsilon N_2} y_2(L) \geq 0 \tag{5.A.1.7}$$

where $Z_j(L^*)$, $j = 1, 2$, are defined in (5.3.6).

If we can determine L^*, we can have $n(0)$ and $n(L)$ from (5.A.1.5). To determine L^*, we substitute (5.A.1.5) and $c_j(\omega) = \xi_j \bar{s}_j y_j(\omega)$ into (5.1.17) to get

$$C_1 = \frac{(1 - Z_1^{2+\lambda})\xi_1 \bar{s}_1 y_1^2}{(2 + \lambda)\upsilon y_1^*(L^*)}, \quad C_2 = \frac{(Z_2^{2+\lambda} - 1)\xi_2 \bar{s}_2 y_2^2(L)}{(2 + \lambda)\upsilon y_2^*(L^*)} \tag{5.A.1.8}$$

in which we use (5.A.1.7). We can thus express C_j ($j = 1, 2$) as functions of y_j ($j = 1, 2$) and L^*.

Substituting (5.A.1.1) into (5.1.18) yields

$$K_i + (C_1 + C_2)\frac{\alpha_h \eta_2 K_i}{\alpha \xi_2 F} = K_1 + K_2. \tag{5.A.1.9}$$

Substituting (5.3.1) into (5.1.12), we have

$$C_1 + C_2 = F - \delta K_1 - \delta K_2. \tag{5.A.1.10}$$

From (5.A.1.9) and (5.A.1.10) we have

$$(1 + u_0)K_i = \left(1 + \frac{\delta u_0 K_i}{F}\right)(K_1 + K_2). \tag{5.A.1.11}$$

From (5.3.3), we have

$$K_1 = \frac{\beta N_1 K_i}{(\delta K_i / s_1 F - \alpha)N^*}, \quad K_2 = \frac{\beta z N_2 K_i}{(\delta K_i / s_2 F - \alpha)N^*}. \tag{5.A.1.12}$$

As K_1 and K_2 are meaningful only when they are positive, we should require

$$\frac{K_i}{F} > K_0 \equiv \max\left\{\frac{\alpha s_j}{\delta}, j = 1, 2\right\}. \qquad (5.A.1.13)$$

Adding the two equations in (5.A.1.12) yields

$$K_1 + K_2 = \left[\frac{N_1}{\delta K_i / s_1 F - \alpha} + \frac{zN_2}{\delta K_i / s_2 F - \alpha}\right] \frac{\beta K_i F}{N^*}. \qquad (5.A.1.14)$$

Substituting (5.A.1.12) into (5.A.1.11) yields (5.3.4). It is easy to check that

$$\Phi(K_0) > 0, \quad \Phi(\infty) = u_0(s_1 N_1 + z s_2 N_2) - \frac{1 + u_0}{\beta} N^* < 0 \qquad (5.A.1.15)$$

in which we use $N^* = N_1 + zN_2$ and $0 < \beta, s_1, s_2 < 1$. From (5.A.1.15), we see that there is at least one positive K^*, $K_0 < K^* < \infty$, such that $\Phi(K^*) = 0$. Taking derivatives of (5.A.1.15) with respect to K^* yields

$$\frac{d\Phi}{dK^*} = -\left(\frac{1}{s_1} + \alpha u_0\right)\delta N_1 S_1^2 - \left(\frac{1}{s_2} + \alpha u_0\right)\delta z N_2 S_2^2 < 0. \qquad (5.A.1.16)$$

From (5.A.1.15) and (5.A.1.16), we see that $\Phi(K^*) = 0$ has a unique positive solution as a function of the parameters in the system. The capital stock employed by the industrial sector is given by

$$K_i = K^{*1/\beta} N^*. \qquad (5.A.1.17)$$

From (5.A.1.12), we directly solve the capital stocks, K_1 and K_2, owned by the two groups. We have thus proved the lemma.

A.5.2 Proving Lemma 5.3.2

The Jacobian at equilibrium is given by

5 Dynamic Urban Pattern Formation with Heterogeneous Population

$$J = \begin{pmatrix} s_1 \dfrac{\partial Y_1}{\partial K_1} - \delta & s_1 \dfrac{\partial Y_1}{\partial K_2} \\ s_2 \dfrac{\partial Y_2}{\partial K_1} & s_2 \dfrac{\partial Y_2}{\partial K_2} - \delta \end{pmatrix} \qquad (5.A.2.1)$$

in which

$$\dfrac{\partial Y_1}{\partial K_1} = \left(1 - \dfrac{\beta u_1 K_1}{K_i} + \dfrac{\beta u_1 N_1}{N^*}\right) \dfrac{\alpha F}{K_i},$$

$$\dfrac{\partial Y_1}{\partial K_2} = \left(\dfrac{N_1}{N^*} - \dfrac{K_1}{K_i}\right) \dfrac{\alpha \beta u_2 F}{K_i},$$

$$\dfrac{\partial Y_2}{\partial K_1} = \left(\dfrac{N_2}{N^*} - \dfrac{K_2}{K_i}\right) \dfrac{\alpha \beta u_1 z F}{K_i},$$

$$\dfrac{\partial Y_2}{\partial K_2} = \left(1 - \dfrac{\beta u_2 K_2}{K_i} + \dfrac{\beta u_2 z N_1}{N^*}\right) \dfrac{\alpha F}{K_i} \qquad (5.A.2.2)$$

where we use (5.3.2). In (5.A.2.2), we calculate $\partial K_i / \partial K_j = u_j$, $j = 1, 2$, by using the following equation

$$K_i = u_1 K_1 + u_2 K_2 \qquad (5.A.2.3)$$

in which

$$u_1 \equiv \dfrac{1 + \alpha s_1 u_0}{N^* + u_0 N^* - \beta s_1 u_0 N_1 - \beta z s_2 u_0 N_2} N^* > 0,$$

$$u_2 \equiv \dfrac{1 + \alpha s_2 u_0}{N^* + u_0 N^* - \beta s_1 u_0 N_1 - \beta z s_2 u_0 N_2} N^* > 0. \qquad (5.A.2.4)$$

The equation, (5.A.2.3), is determined by (5.1.12), (5.3.2) and (5.3.9).

We know that if

$$\Delta_1 \equiv s_1 \dfrac{\partial Y_1}{\partial K_1} + s_2 \dfrac{\partial Y_2}{\partial K_2} - 2\delta < 0,$$

$$\Delta_2 \equiv \left(s_1 \frac{\partial Y_1}{\partial K_1} - \delta\right)\left(s_2 \frac{\partial Y_2}{\partial K_2} - \delta\right) - s_1 s_2 \frac{\partial Y_2}{\partial K_1}\frac{\partial Y_1}{\partial K_2} > 0 \quad (5.A.2.5)$$

then the equilibrium is stable. From (5.A.2.2) and (5.A.2.5), we have

$$\frac{\Delta_1}{\beta F} = -s_1 \left[\frac{\alpha u_1 K_1}{K_i^2} + \frac{\beta u_1 K_1 + u_2 K_2}{K_1 K_i N^*} N_1\right]$$

$$- s_2 \left[\frac{\alpha u_2 K_2}{K_i^2} + \frac{u_1 K_1 + \beta u_2 K_2}{K_2 K_i N^*} z N_2\right] < 0,$$

$$\frac{K_i^2 \Delta_2}{s_1 s_2 \beta^2 F^2} = \alpha^2 z u_1 u_2 \left[\frac{N_1}{N^*} - \frac{K_1}{K_i}\right]\left[\frac{N_2}{N^*} - \frac{K_2}{K_i}\right] +$$

$$\left[\frac{\alpha u_1 K_1}{K_i} + \frac{\beta u_1 K_1 + u_2 K_2}{K_1 N^*} N_1\right]\left[\frac{\alpha u_2 K_2}{K_i} + \frac{u_1 K_1 + \beta u_2 K_2}{K_2 N^*} z N_2\right]$$

(5.A.2.5)

where we use (5.A.2.3) and

$$\delta = \left(\frac{\alpha}{K_i} + \frac{\beta N_1}{K_1 N^*}\right) s_1 F = \left(\frac{\alpha}{K_i} + \frac{z \beta N_2}{K_2 N^*}\right) s_2 F.$$

Summarizing the above discussions, we have completed the proof of the lemma.

6 Two-Group Spatial Structures with Capital and Knowledge

This chapter introduces endogenous knowledge into the model with heterogeneous population proposed in the previous chapter. Whether economic development will result in the divergence or convergence of income and wealth distribution among various groups of people is one of the main concerns of the classical economists. But only a few theoretical dynamic models explicitly deal with the issues concerning the co-existence of multiple groups. Issues related to the co-existence, separation and stability of multiple income classes are well raised in urban economics within static frameworks of exogenous incomes (e.g., Yinger, 1976, Yellin, 1974, Schare, 1976, Zhang, 1989a). This chapter examines the effects of differences in knowledge utilization and savings behavior between the two groups upon long-run growth and wealth and income distribution within the framework of spatial economics. This chapter is based on Zhang (1993c).

6.1 The Spatial Economy with Heterogeneous Population

The economic production in this chapter is the same as that in the one-sector models in Chapter 2. The economic system is an isolated state located in a round island (with a fixed radius). The system is geographically similar to that of the model in Chapter 5. Similarly to the model in the previous chapter, we classify the population into two groups and neglect any possibility of group transformation. We assume that there is no prejudice in the labor market. Here, by justice in the labor market we mean that any worker is paid according to the marginal value of 'qualified labor'. Markets are characterized by perfect competition. It is not only working hours that determine how much one is paid. Wage rates are determined by a combination of working time and human capital. For simplicity, we assume that the human capital of the workers from the same group is identical and there are differences in human capital between the two groups. As the labor market is perfectly competitive, the wage rate per unit of 'qualified labor' is identical for all the workers in the system.

Production and capital accumulation
We assume that production is carried out by combining capital and the qualified labor force. Similarly to the neoclassical one-sector growth model, we assume that the

6.1 The Spatial Economy with Heterogeneous Population

production is either invested or consumed. We select the commodity to serve as numeraire and introduce

N_j — the fixed population of group j, $j = 1, 2$;

$N_j^*(t)$ — the qualified labor force (to be defined) of group j;

$K_j(t)$ — capital stocks owned by group j, $j = 1, 2$.

$w_j(t)$ and $r(t)$ — the wage rate of group j and interest rate, respectively;

$F(t)$ — output at time t.

The total qualified labor force $N^*(t)$ and the total capital stock $K(t)$ of the country are defined by

$$N^* = N_1 + N_2, \quad K = K_1 + K_2. \tag{6.1.1}$$

We assume that the labor force and capital are always fully employed.

In order to describe the relationship among N_j and N_j^*, we introduce the variable of the knowledge stock $Z(t)$ of the economic system. Here, we treat the knowledge stock $Z(t)$ as a 'public good' in the sense that learning of a group from the 'knowledge reservoir' $Z(t)$ will not affect that of the other group. Due to differences in the factors, such as attitudes toward learning, social positions, accessibilities to education systems, and cultural traditions, the two groups may utilize knowledge differently. For simplicity of analysis, we specify the following relationship among N_j and N_j^*

$$N_j^* = Z^{m_j} N_j, \quad j = 1, 2 \tag{6.1.2}$$

in which m_j is a positive parameter. We call m_j the knowledge utilization efficiency parameter of group j. It is a measurement of the productivity of group j in the production function. The specified functional form implies that an increase in the knowledge stock will increase human capital of the two groups, but in different proportions. Here, we neglect the possible impact of education, training and other costly learning efforts upon N_j^*. At this initial stage we neglect any possible impact of education, training, wealth accumulation or other factors upon the knowledge utilization parameters, m_j.

6 Two-Group Spatial Structures with Capital and Knowledge

The producer employs the production factors, maximizing

$$F(K, N_1^* + N_2^*) - rK - w_1 N_1 - w_2 N_2. \qquad (6.1.3)$$

Here, we specify $F(K, N^*)$ in the following form

$$F(K, N^*) = K^\alpha N^{*\beta}, \quad \alpha + \beta = 1, \quad \alpha, \beta > 0. \qquad (6.1.4)$$

The profit maximization yields the following conditions

$$r = \frac{\alpha F}{K}, \quad w_j = \frac{\beta Z^{m_j} F}{N^*}, \quad j = 1, 2. \qquad (6.1.5)$$

From (6.2.5), we directly obtain the following important result.

Lemma 6.1.1.
For any given knowledge stock $Z(t)$ at any point of time, the ratio of the wage rates w_1/w_2 between groups 1 and 2 is only related to the difference between the knowledge utilization efficiency parameters $m_1 - m_2$ between the two groups.

When $m_1 = m_2$, the two groups have an identical wage rate. That is, the difference in wages is due to the difference in human capital in economic activities in our model. Define w by

$$w \equiv \frac{w_1}{Z^{m_1}} = \frac{w_2}{Z^{m_2}}. \qquad (6.1.6)$$

We interpret w as the wage rate per unit of qualified labor force. Equation (6.1.6) implies that there is no group prejudice such as racial discrimination in the labor market. The income of group j, Y_j, $j = 1, 2$, is given by

$$Y_j = rK_j + w_j N_j, \quad j = 1, 2. \qquad (6.1.7)$$

Substituting (6.1.4) and (6.1.5) into (6.1.7) and then adding the two equations, we have

$$Y_1 + Y_2 = F(K, N^*). \qquad (6.1.8)$$

From (6.1.7), the difference $y_1 - y_2$ in incomes per capita is given by

6.1 The Spatial Economy with Heterogeneous Population

$$r(k_1 - k_2) + w_1 - w_2$$

where $y_j = Y_j / N_j$ and $k_j = K_j / N_j$.

The capital accumulation of group j is given by

$$\frac{dK_j}{dt} = s_j Y_j - \delta_k K_j, \quad j = 1, 2 \tag{6.1.9}$$

where δ_k is the depreciation rate of capital, and s_j are the constant savings rates of group j. In the remainder of this chapter, we assume that the two groups have an identical savings rate, i.e., $s = s_1 = s_2$. This is obviously a very strict assumption. As shown soon, this assumption simplifies our analysis, reducing the three-dimensional dynamics into a two-dimensional one.

Adding the two equations in (6.1.9) together with (6.1.8) yields

$$\frac{dK}{dt} = sF(K, N^*) - \delta_k K. \tag{6.1.10}$$

Knowledge accumulation
Similarly to Chapter 3, we propose the following possible dynamics of knowledge

$$\frac{dZ}{dt} = \frac{\tau_1 N_1^* F}{N^* Z^{\varepsilon_1}} + \frac{\tau_2 N_2^* F}{N^* Z^{\varepsilon_2}} - \delta_z Z \tag{6.1.11}$$

in which $\tau_j \, (\geq 0)$, ε_j, and $\delta_z \, (\geq 0)$ are parameters.

Residential pattern
We have described the economic activities at the CBD. We are now concerned with location and residential patterns. First, we introduce

L — the fixed radius, i.e., the distance from the CBD to the boundary of the island;
ω — the distance from the CBD to a point in the residential area;
$R(\omega, t)$ — the land rent at location ω and at time t;
$L_{hj}(\omega, t)$ — the lot size per household of group j, $j = 1, 2$;
$n_j(\omega, t)$ — the residential density of group j ($j = 1, 2$);
$c_j(\omega, t)$ — the consumption level per household of group j; and

$C_j(\omega)$ — the total consumption level of group j, $j = 1, 2$.

We assume that utility levels are dependent upon lot size and consumption levels. For simplicity of analysis, we specify the utility level U_j of group j as follows

$$U_j(\omega,t) = c_j^{\xi_j} L_{hj}^{\xi_j} \tag{6.1.12}$$

where ξ_j, $j = 1, 2$, are positive parameters. It should be noted that it is important to introduce other locational factors such as amenities and infrastructures in the analysis of urban pattern formation (e.g., Diamond and Tolley, 1981, Nijkamp, 1986, Simon and Love, 1990, Vomit, 1991, Blomquist, Berger and Hoehn, 1988).

As the consumption budgets of the two groups are given by $(1 - s)Y_j$, $j = 1, 2$, respectively, the consumer problem is defined by

$$\max U_j \text{ s.t: } c_j(\omega) + R(\omega)L_{hj}(\omega) = y_j(\omega) \tag{8.1.13}$$

where

$$y_j(\omega) \equiv q_i - \Gamma(\omega), \quad q_i(t) \equiv \frac{(1 - s_j)Y_j}{N_j}, \quad j = 1, 2 \tag{6.1.14}$$

and $\Gamma(\omega)$ is the total traveling cost between dwelling site ω and the CBD. We specify $\Gamma(\omega)$ by

$$\Gamma(\omega) = \upsilon_0 \omega^\upsilon, \quad \upsilon_0 > 0, \quad 0 \le \upsilon \le L. \tag{6.1.15}$$

We require $y_j(L) > 0$, i.e.

$$q_j(t) > \Gamma(L), \quad j = 1, 2. \tag{6.1.16}$$

That is, the consumption level of any group is sufficient for traveling from the CBD to the edge of the island.

The unique optimal solution is given by

$$c_j = \frac{y_j}{2}, \quad L_{hj} = \frac{y_j}{2R}, \quad j = 1, 2. \tag{6.1.17}$$

According to the definitions, we have

$$n_j(\omega,t) = \frac{1}{L_{hj}(\omega,t)}, \quad 0 \leq \omega \leq L, \quad j = 1, 2. \quad (6.1.18)$$

The population constraints are given by

$$2\pi \int_0^L n_j(\omega)\omega \, d\omega = N_j, \quad j = 1, 2. \quad (6.1.19)$$

The consumption constraints are given by

$$2\pi \int_0^L c_j(\omega) n_j(\omega)\omega \, d\omega = C_j = (1 - s_j)Y_j, \quad j = 1, 2. \quad (6.1.20)$$

We have thus finished building the model. The framework suggested in this chapter is, from a structural point of view, general in the sense that some well-known models in theoretical economics can be considered as special cases of our model. For instance, if we neglect land, housing, endogenous knowledge and the classification of the population, our model is similar to the neoclassical one-sector model. Adding the classification of the population to the Solow-Swan model, our system is similar to the two-group growth model (Sato, 1966). It is not difficult to see that the basic structure of our system with one-sector and learning by doing is similar to the Arrow one-sector learning by doing model (Arrow, 1962).

6.2 The Temporary Urban Pattern

This section proves the existence of a unique residential pattern for given stocks of capital and knowledge at any point of time.

As the households of the same group have an identical utility function, the utility level U_j is identical over space within the same group. Utilizing $U_j(\omega_1,t) = U_j(\omega_2,t)$, for any ω_1 and ω_2, $0 \leq \omega_1, \omega_2 \leq L$, $j = 1, 2$, together with (6.1.12) and (6.1.17), we have

$$\frac{R(\omega_1)}{R(\omega_2)} = \left[\frac{y_j(\omega_1)}{y_j(\omega_2)}\right]^2, \quad j = 1, 2 \quad (6.2.1)$$

106   6 Two-Group Spatial Structures with Capital and Knowledge

at any point of time. From (6.2.1), we see that if $q_1 = q_2$, the two groups may co-exist at location ω. If the two groups do not have identical incomes per capita, we conclude that

$$\frac{y_1(\omega_1)}{y_1(\omega_2)} = \frac{y_2(\omega_1)}{y_2(\omega_2)}$$

cannot hold. This implies that the two groups cannot co-exist at any two points in the whole urban area. This further implies that there is a distance $L^*(t)$, $0 < L^*(t) < L$ such that the two groups are separately located within $\omega \in [0, L^*)$ and $\omega \in [L^*, L]$. However, we cannot determine which group will locate nearer to the CBD without further constraints on decision making about location. To solve the problem, as in the previous chapter, we introduce the following well-known rule in urban economics.

The rule of determining groups' location
If the rent function of group k is steeper than that of group j at the intersection, the equilibrium location of group k is closer to the CBD than that of group j.

This rule is discussed in the literature of urban economics (Fujita, 1989). Utilizing

$$R(L^*, U_j^*) = \frac{y_j^2}{4p_s U_j^{*1/\xi_j}}$$

where U_j^* is the utility level evaluated at $\omega = L^*$, we directly have

$$\frac{\partial R(L^*, U_j^*)}{\partial \omega_j} = -\frac{2\upsilon_0 \upsilon L^{*\upsilon-1} R(L^*)}{y_j(L^*)} \qquad (6.2.2)$$

where $\partial R(L^*, U_j^*)/\partial \omega_j$ is the partial derivative of $R(L^*, U_j^*)$ with respect to ω for group j (with U_j^* fixed). We have

$$-\frac{\partial R}{\partial \omega_k} > -\frac{\partial R}{\partial \omega_j} \quad \text{when} \quad y_k(L^*) < y_j(L^*).$$

This implies that the group with the higher income per capita locates farther from the CBD than the other group.

Lemma 6.1.1.
Assume that the incomes per capita are not identical for both groups. Then, at any point of time, according to the rule of determining groups' location, the households of each group form a concentric ring around the CBD and the group with higher income per capita locates farther from the CBD than the group with the lower one. Moreover, from (6.2.1), $R(\omega)$ is continuously decreasing up to the urban fringe L.

In the remainder of this chapter, for convenience of discussion we assume that $q_1 < q_2$, i.e., $Y_2/N_2 > Y_1/N_1$, at any point of time. Under this requirement group 2 locates farther from the CBD than group. It can be easily seen that $Y_2/N_2 > Y_1/N_1$ at any point of time is held if $m_2 > m_1$ and $K_2/N_2 > K_1/N_1$ at $t = 0$.

Assumption 6.2.1.
In the remainder of this chapter, we assume that group 2 has a greater knowledge utilization efficiency than group 1, i.e., $m_2 > m_1$, and the initial capital stock per capita of group 2 is greater than that of group 1.

This assumption guarantees that group 2 is always located further from the CBD than group 1. In Appendix A.6.1, we show that there is a unique urban structure for any given K and Z at any point of time.

Lemma 6.2.2.
Let $q_j(t) > \Gamma(L)$ be satisfied at any point of time. Then for any given $K(t)$ and $Z(t)$ there is a unique intersection, $L^*(t), 0 < L^* < L$, between the two groups at any point of time. The land rent and residential distribution at any point of time are uniquely given by

$$R(L^*) = \frac{q_1 L^{*2} - \upsilon^* L^{*\upsilon+2}}{2n_{01} y_1(L^*)},$$

$$R(\omega) = R(L^*)\left[\frac{y_1(\omega)}{y_1(L^*)}\right]^2,$$

$$n_1(\omega) = \frac{2R(\omega)}{y_1(\omega)}, \quad 0 \leq \omega \leq L^*, \tag{6.2.3}$$

$$R(L) = R(L^*)\left[\frac{y_2(L)}{y_2(L^*)}\right]^2,$$

108   6 Two-Group Spatial Structures with Capital and Knowledge

$$R(\omega) = R(L^*)\left[\frac{y_2(\omega)}{y_2(L^*)}\right]^2,$$

$$n_2(\omega) = \frac{2R(\omega)}{y_2(\omega)}, \quad L^* \leq \omega \leq L. \tag{6.2.4}$$

The above lemma is proved in Appendix 6.A.1. The lemma guarantees the existence of an urban structure for any given income of either group at any point of time. We now examine how capital and knowledge are determined as time passes.

6.3 The Long-Run Equilibria and Stability Conditions

This section is concerned with the dynamic properties of the dynamic system. First, by (6.1.10) and (6.1.11), we can rewrite the three-dimensional dynamics in terms of K_1, K, and Z as follows

$$\frac{dK_1}{dt} = sY_1(K_1, K, Z) - \delta_k K_1,$$

$$\frac{dK}{dt} = sK^\alpha N^{*\beta} - \delta_k K,$$

$$\frac{dZ}{dt} = \left[\frac{\tau_1^* N_1^*}{Z^{\varepsilon_1}} + \frac{\tau_2^* N_2^*}{Z^{\varepsilon_2}}\right]\left(\frac{K}{N^*}\right)^\alpha \beta - \delta_z Z \tag{6.3.1}$$

where $\tau_j^* \equiv \tau_j / \beta$, $j = 1, 2$, and

$$Y_1(K_1, K, Z) \equiv \left(\frac{\alpha K_1}{K} + \frac{\beta N_1^*}{N^*}\right) F(K, N^*). \tag{6.3.2}$$

As the last two differential equations for K and Z in (6.3.1) are independent of the first one for K_1, we may carry out a dynamic analysis of the last two equations, disregarding the equation for K_1. We can directly check that for any given K and Z, $dK_1/dt = sY_1 - \delta_k K_1$ has a unique stable positive equilibrium. We thus can reduce one dimension of the dynamic system.

It is proved that the dynamic properties of the system are determined by the following two parameters

$$x_j = m_j - \varepsilon_j - 1, \quad j = 1, 2. \tag{6.3.3}$$

The following proposition is proved in the appendix of this chapter.

Proposition 6.3.1.
(i) If $x_j < 0$, $j = 1, 2$, the system has a unique stable equilibrium; (i) If $x_j > 0$, $j = 1, 2$, the system has a unique unstable equilibrium; and (iii) If $x_1 > 0$ and $x_2 < 0$ ($x_1 < 0$ and $x_2 > 0$), the system has either two positive equilibria or no equilibrium. When the system has two equilibria, the one with low values of K and Z is stable; the other one is unstable.

We may interpret the two parameters x_j as in Chapter 5.

Economic interpretations of Proposition 6.3.1.
If the knowledge utilization and creation of the two groups exhibit decreasing (increasing) return to scale effects in the system, the system has a unique stable (unstable) equilibrium; If the knowledge utilization and creation of one group exhibits decreasing return to scale effects in the system and the other one exhibits increasing scale effects, the system has two equilibria. The one with lower values of K and Z is stable, while the other one is unstable.

6.4 The Impact of Creativity

In this and the following sections, we examine the impact of changes in the creativity parameters τ_j and the population N_j on economic growth and urban structure. We carry out comparative statics analysis under presumed stability, i.e.

$$\Phi^* \equiv -\frac{(x_1\Phi_1 + x_2\Phi_2)}{Z} > 0.$$

When the system has two equilibria, our conclusions are valid for the equilibrium with lower values of K and Z. It can be seen that if the equilibrium is unstable, i.e., $\Phi^* < 0$, we may get the opposite results to those we will have under presumed stability.

Taking derivatives of (6.A.2.2) and (6.A.2.3) with respect to τ_j yields

$$\frac{dK}{d\tau_j} = \left[\frac{(m_1 N_1^* + m_2 N_2^*)K}{N^* Z}\right]\frac{dZ}{d\tau_j} > 0,$$

$$\Phi^* \frac{dZ}{d\tau_j} = \frac{\Phi_j}{\tau_j} > 0, \quad j = 1, 2. \tag{6.4.1}$$

An increase in the creativity of any group will expand the total stocks of capital and knowledge of the society. From $sY_j = \delta_k K_j$, $sF = \delta_k K$, (6.1.5), and (6.1.7), we have $K_j = K N_j^* / N^*$. Hence,

$$\frac{d(K_1 / K_2)}{d\tau_j} = \frac{d(Y_1 / Y_2)}{d\tau_j} = \left[\frac{(m_1 - m_2)K_1}{ZK_2}\right]\frac{dZ}{d\tau_j},$$

$$\frac{dK_j}{d\tau_j} = \frac{m_j K_j}{Z}\frac{dZ}{d\tau_j} > 0, \quad \frac{dY_j}{d\tau_j} > 0,$$

$$\frac{dk_j}{d\tau_j} > 0, \quad \frac{dy_j}{d\tau_j} > 0, \quad \frac{dq_j}{d\tau_j} > 0 \tag{6.4.2}$$

are held. By (6.1.4), (6.1.5), and (6.1.7), we have

$$\frac{dF}{d\tau_j} > 0, \quad \frac{dr}{d\tau_j} = 0, \quad \frac{dw_j}{d\tau_j} > 0,$$

$$\frac{d(w_1 / w_2)}{d\tau_j} = \left[\frac{(m_1 - m_2)w_1}{Zw_2}\right]\frac{dZ}{d\tau_j}. \tag{6.4.3}$$

All the above conclusions are easy to interpret. We now examine the impact upon the intersection between the two groups, residential density and rent distribution. Taking the derivative of (6.A.1.5) with respect to τ_j yields

$$-M'\frac{dL^*}{d\tau_j} = \Gamma_1 \frac{dY_1}{d\tau_j} + [v_0(L^v - L^{*v}) + \Gamma_2 - y_2(L)]\Gamma_3 \frac{dY_2}{d\tau_j} \tag{6.4.4}$$

in which $M' (< 0)$ is defined in (6.A.1.6), $dY_1 / d\tau_j > 0$, $dY_2 / d\tau_j > 0$, and

6.4 The Impact of Creativity

$$\Gamma_1 \equiv \frac{\upsilon_0 \upsilon q_1 L^{*\upsilon+2}}{Y_1 y_1(L^*)(2+\upsilon)} > 0,$$

$$\Gamma_2 \equiv \frac{(L^2 - L^{*2}) y_2(L) y_2(L^*)}{q_2(L^2 - L^{*2}) - \xi_0(L^{\upsilon+2} - L^{*\upsilon+2})} > 0,$$

$$\Gamma_3 \equiv \frac{q_2 L^{*2}(q_1 - \xi_0 L^{*\upsilon})}{y_2(L) y_2(L^*) Y_2} > 0. \tag{6.4.5}$$

We see that $dL^*/d\tau_j$ may be either positive or negative. Since the incomes of the two groups are increased due to the improved creativity of either group, each group tends to expand residential space. As the total urban size is fixed, the intersection may be moved outward or onwards due to the interdependence of various forces.

Taking derivatives of $R(L^*)$ in (6.2.3) yields

$$\frac{2n_{01} y(L^*) y_1(L^*)}{L^{*2}} \frac{dR(L^*)}{d\tau_j} = -\frac{\upsilon_0 \upsilon q_1 L^{*\upsilon}}{(2+\upsilon) Y_1} \frac{dY_1}{d\tau_j}$$

$$\left[\frac{\upsilon^2 \upsilon_0 q_1 L^{*\upsilon-1}}{2+\upsilon} + \frac{(q_1 - \xi_0 L^{*\upsilon}) 2 y(L^*)}{L^*} \right] \frac{dL^*}{d\tau_j}. \tag{6.4.6}$$

Although the sign of $dR(L^*)/d\tau_j$ is generally uncertain since $dL^*/d\tau_j$ may be either positive or negative, it is negative in the case of $dL^*/d\tau_j < 0$. That is, if the intersection is moved onwards, the land rent at the intersection tends to fall as the creativity of either group is increased.

From (6.2.3), we directly get the impact of changes in τ_j upon R and n_j in group 1's residential zone as follows

$$\frac{1}{R(\omega)} \frac{dR(\omega)}{d\tau_j} = \left[\frac{q_1 \upsilon_0 (\omega^\upsilon - L^{*\upsilon})}{Y_1} \frac{dY_1}{d\tau_j} - \upsilon_0 \upsilon L^{*\upsilon-1} \frac{dL^*}{d\tau_j} \right]$$

$$\frac{2}{y_1(\omega) y_1(L^*)} + \frac{1}{R(L^*)} \frac{dR(L^*)}{d\tau_j},$$

$$\frac{1}{n_1(\omega)}\frac{dn_1(\omega)}{d\tau_j} = \frac{1}{R(\omega)}\frac{dR(\omega)}{d\tau_j} - \frac{q_1}{y_1(\omega)Y_1}\frac{dY_1}{d\tau_j}, \quad 0 \le \omega < L^*.$$

(6.4.7)

We see that $dR(\omega)/d\tau_j$ and $dn_1(\omega)/d\tau_j$ may be either positive or negative. In the case of $dR(\omega)/d\tau_j < 0$, the residential density at location ω surely falls. But when $dR(\omega)/d\tau_j > 0$, the residential density may either fall or rise.

Taking derivatives of (6.2.4) with respect to τ_j yields

$$\frac{1}{R(L)}\frac{dR(L)}{d\tau_j} = \left[\frac{q_2 \upsilon_0 (L^\upsilon - L^{*\upsilon})}{Y_2}\frac{dY_2}{d\tau_j} - \upsilon_0 \upsilon L^{*\upsilon-1}\frac{dL^*}{d\tau_j}\right]$$

$$\frac{2}{y_2(L)y_2(L^*)} + \frac{1}{R(L^*)}\frac{dR(L^*)}{d\tau_j},$$

$$\frac{1}{R(\omega)}\frac{dR(\omega)}{d\tau_j} = \frac{2q_2 \upsilon_0 (\omega^\upsilon - L^\upsilon)}{y_2(L)y_1(\omega)Y_2}\frac{dY_2}{d\tau_j} + \frac{1}{R(L)}\frac{dR(L)}{d\tau_j},$$

$$\frac{1}{n_2(\omega)}\frac{dn_2(\omega)}{d\tau_j} = \frac{1}{R(\omega)}\frac{dR(\omega)}{d\tau_j} - \frac{q_2}{y_2(\omega)Y_2}\frac{dY_2}{d\tau_j}, \quad L^* \le \omega < L.$$

(6.4.8)

It is also difficult to explicitly judge the impact upon R and n_2 for $L^* \le \omega < L$. But it should be remarked that even if $dL^*/d\tau_j < 0$, $dR(L)/d\tau_j$ may be either positive or negative.

Summarizing the above analysis, we have the following proposition.

Proposition 6.4.1.
Assume that the equilibrium under consideration be stable, i.e., $\Phi^* > 0$, and that group 2 utilizes knowledge more effectively than group 1, i.e., $m_2 > m_1$. Then, an increase in the creativity parameter τ_j, $j = 1, 2$, of either group has the following long-run impact upon the system
i) the total output and the total stocks of capital and knowledge are increased, i.e., $dF/d\tau_j$, $dK/d\tau_j > 0$ and $dZ/d\tau_j > 0$;

ii) the capital stock per capita k_j, the income per capita y_j, and the wage rate w_j, of each group are increased;

iii) the ratios of the capital stocks K_1/K_2, incomes Y_1/Y_2 and wage rates w_1/w_2 between groups 1 and 2 are decreased; and

iv) the impact upon the intersection L^*, the residential density distribution and the rent distribution may be either positive or negative.

6.5 The Impact of the Population

From (6.A.2.2) and (6.A.2.3), we have

$$\frac{1}{K}\frac{dK}{dN_j} = \left(\frac{m_1 N_1^* + m_2 N_2^*}{N^* Z}\right)\frac{dZ}{dN_j} + \frac{N_j^*}{ZN_j N^*} > 0,$$

$$\Phi^* \frac{dZ}{dN_j} = \frac{\Phi_j}{N_j} > 0, \quad j = 1, 2. \tag{6.5.1}$$

An increase in the population of either group will expand the capital and knowledge stocks of the society. By $K_j = KN_j^*/N^*$, (6.1.4), (6.1.5), and (6.1.7), we have

$$\frac{dK_j}{dN_j} > 0, \quad \frac{dY_j}{dN_j} > 0,$$

$$\frac{dF}{dN_j} > 0, \quad \frac{dr}{dN_j} = 0, \quad \frac{dw_j}{dN_j} > 0,$$

$$\frac{d(w_1/w_2)}{dN_j} = \left[\frac{(m_1 - m_2)w_1}{Zw_2}\right]\frac{dZ}{dN_j},$$

$$\frac{d(K_1/K_2)}{dN_1} = \frac{d(Y_1/Y_2)}{dN_1} = \left[\frac{(m_1 - m_2)K_1}{ZK_2}\right]\frac{dZ}{dN_1} + \frac{K_1}{N_1 K_2},$$

$$\frac{d(K_1/K_2)}{dN_2} = \frac{d(Y_1/Y_2)}{dN_2} = \left[\frac{(m_1 - m_2)K_1}{ZK_2}\right]\frac{dZ}{dN_2} - \frac{K_1}{N_2 K_2}. \tag{6.5.2}$$

114   6 Two-Group Spatial Structures with Capital and Knowledge

Changes in N_j have similar effects upon K_j, K_1/K_2, K, Z, F, Y, w_j, Y_1/Y_2, r, w_1/w_2, to the effects of changes in τ_j. Utilizing (6.A.1.5), (6.2.3) and (6.2.4), we can also give the impact upon the urban pattern. Since the effects are similar to the corresponding ones for changes in τ_j, we omit any presentation of the results.

6.6 On Urban Evolution with Heterogeneous Population

This chapter proposed an economic growth model with location choice and a land use pattern. The model may be considered as a synthesis of the Alonso model with multiple-income groups and the one-sector growth model with endogenous knowledge. The model examined the dynamic interdependence between knowledge utilization, creativity, transportation conditions, savings behavior, location choice and residential patterns in a two-group economy with justice in the labor market. It was proved that the system may have a unique equilibrium or multiple equilibria and each equilibrium may exhibit stability or instability, depending upon the knowledge utilization and creation characteristics of the two groups. We also examined the impact of changes in the population and the knowledge creation efficiency of the two groups upon long-run growth, wealth distribution and residential structure.

As the model is based upon two well-established modeling frameworks (in growth theory and urban economics), it is conceptually easy to extend this model along the lines upon which those two frameworks have been extended and generalized. For instance, we may consider an economy consisting of multiple groups with a family structure (e.g., Beckmann, 1973), of multiple groups with prejudice (e.g., Yellin, 1974), of urban externalities (e.g., Henderson, 1985, Bell, 1991, Roback, 1982, Sivitanidou and Wheaton, 1992), of multiple production sectors (e.g., Zhang, 2000), of multiple nations (e.g., Rauch, 1991, Zhang, 1992b, 1995b) and so on. Indeed, these extensions may be difficult in analytical terms.

Appendix

A.6.1 Proving Lemma 6.2.2

We now prove lemma 6.2.2. First, from (6.2.1), we can express the land rent distribution in terms of land rents at $\omega = L^*$ and $\omega = L$ as follows

$$R(\omega) = \left[\frac{y_1(\omega)}{y_1(L^*)}\right]^2 R(L^*), \quad 0 \le \omega < L^*,$$

$$R(\omega) = \left[\frac{y_2(\omega)}{y_2(L^*)}\right]^2 R(L), \quad L^* \le \omega < L. \qquad (6.A.1.1)$$

From (6.1.10) and (6.1.11), we have

$$n_1(\omega) = \frac{1}{L_{h1}(\omega)} = \frac{2R(\omega)}{y_1(\omega)}, \quad 0 \le \omega < L^*,$$

$$n_2(\omega) = \frac{1}{L_{h2}(\omega)} = \frac{2R(\omega)}{y_2(\omega)}, \quad L^* \le \omega \le L. \qquad (6.A.1.2)$$

Substituting (6.A.1.2) together with (6.A.2.5), respectively, into (6.1.12) yields

$$M_1(L^*, M) \equiv q_1 L^{*2} - \xi_0 L^{*v+2} - \frac{2n_{01} y_1(L^*)}{R(L^*)} = 0,$$

$$M_2(L^*, M) \equiv q_2(L^2 - L^{*2}) - \xi_0(L^{*v+2} - L^{v+2}) - \frac{2n_{02} y_2(L)}{R(L)} = 0$$

$$(6.A.1.3)$$

in which $n_{0j} \equiv N_j / 4\pi$ and $\xi_0 \equiv 2v_0 /(2 + v)$. We now show that from (6.A.1.1) and (6.A.1.3), we can uniquely determine L^* and $R(\omega)$ for $0 \le \omega \le L$.

From $R(L^*)/R(L) = [y_2(L^*)/y_2(L)]^2$ and $M_2 = 0$ in (6.A.1.3), we have

$$R(L^*) = \frac{2n_{02} y_2^2(L^*)}{[q_2(L^2 - L^{*2}) - \xi_0(L^{*v+2} - L^{v+2})] y_2^2(L)}. \qquad (6.A.1.4)$$

Substituting (6.A.1.4) into $M_1 = 0$ yields

$$M(L^*) \equiv 2n_{01} y_1(L^*) y_2(L) \frac{q_2(L^2 - L^{*2}) - \xi_0(L^{*v+2} - L^{v+2})}{n_{02} y_2^2(L^*)}$$

$$- (q_1 - \xi_0 L^{*v}) L^{*2} = 0. \qquad (6.A.1.5)$$

We now prove that $M(L^*) = 0$, $0 < L^* < L$, has a unique positive solution. As $M(0) > 0$ and $M(L) < 0$, the proof is finished if we can show $M'(L^*) < 0$. As

$$M'(L^*) = -L^{*\upsilon-1}M_0(L^*) - \frac{2n_{01}y_1(L^*)y_2(L)}{n_{02}y_2(L^*)}$$
$$-\frac{L^{*\upsilon-1}y_1(L^*)M_0(L^*)}{y_2(L^*)} - 2L^*y_1(L^*) \qquad (6.A.1.6)$$

where

$$M_0(L^*) \equiv \upsilon_0\upsilon n_{01}y_2(L)\frac{q_2(L^2 - L^{*2}) - \xi_0(L^{*\upsilon+2} - L^{\upsilon+2})}{n_{02}y_2^2(L^*)},$$
$$0 \leq L^* \leq L.$$

Since $M_0(L^*) \geq 0$, $0 \leq L^* \leq L$, $M'(L^*) < 0$, $0 \leq L^* \leq L$.

Summarizing the above discussion, we completed the proof of Lemma 6.2.2.

A.6.2 Proving Proposition 6.3.1

A long-run equilibrium of the system is given by

$$sK^{\alpha}N^{*\beta} = \delta_k K,$$
$$\left[\frac{\tau_1^*N_1^*}{Z^{\varepsilon_1}} + \frac{\tau_2^*N_2^*}{Z^{\varepsilon_2}}\right]\left(\frac{K}{N^*}\right)^{\alpha}\beta = \delta_z Z. \qquad (6.A.2.1)$$

From $sK^{\alpha}N^{*\beta} = \delta_k K$, we have

$$K = \left(\frac{s}{\delta_k}\right)^{1/\beta} N^*. \qquad (6.A.2.2)$$

Substituting (6.A.2.2) into the second equation in (6.A.2.1) yields

$$\Phi(Z) \equiv \Phi_1(Z) + \Phi_2(Z) - \delta_z = 0 \qquad (6.A.2.3)$$

in which

$$\Phi_j(Z) \equiv \left(\frac{s}{\delta_k}\right)^{\alpha/\beta} \beta\tau_j^* N_j Z^{x_j}. \qquad (6.A.2.4)$$

Similarly to the proof in Appendix A.5.2 in Chapter 5, we can prove that $\Phi(Z) = 0$ has a unique positive solution in the case of $x_j > 0$ ($x_j < 0$) for $j = 1, 2$. In the case of $x_1 > 0$ and $x_2 < 0$ (or $x_1 < 0$ or $x_2 > 0$), the system has either two solutions or no solution. The two eigenvalues ϕ_j are determined by

$$\phi^2 + \left[\delta_k\beta + \alpha n^*(\Phi_1 + \Phi_2) - x_1\Phi_1 - x_2\Phi_2\right]\phi \\ - (x_1\Phi_1 + x_2\Phi_2)\delta_k\beta = 0 \qquad (6.A.2.5)$$

where

$$n \equiv \frac{m_1 N_1^* + m_2 N_2^*}{N^*}.$$

The system is stable if

$$\delta_k\beta + \alpha n^*(\Phi_1 + \Phi_2) - x_1\Phi_1 - x_2\Phi_2 > 0, \quad x_1\Phi_1 + x_2\Phi_2 < 0.$$

The first inequality holds if $x_1\Phi_1 + x_2\Phi_2 < 0$. We thus conclude that if $x_1\Phi_1 + x_2\Phi_2$ is positive, the equilibrium is unstable; if it is negative, the equilibrium is stable. Summarizing the above analytical results, we completed the proof of Proposition 6.3.1.

7 Urban Growth and Pattern Formation with Preference Change

The main issue in this chapter is preference change. Although preferences for leisure time, housing, food, clothes, traveling and so on are varied over time, only a few urban models explicitly deal with endogenous preference changes (Beckmann, 1990, Kobayashi, Zhang, and Yoshikawa, 1986, 1989). There are dynamic interactions between preference change, production structure and residential location. This chapter constructs a dynamics of preference within the frameworks of growth theory and location theory. We model a dynamic interdependence between urban structure, capital accumulation and preference change in a perfectly competitive economy. The chapter is organized as follows. Section 7.1 defines the model with capital accumulation and preference change. Section 7.2 provides conditions for existence of equilibria and stability. Section 7.3 analyzes the impact of changes in human capital on the equilibrium economic structure. Section 7.5 concludes the chapter. The appendix proves Lemma 7.2.1. This chapter is based on Zhang (1996c, 1998c).

7.1 The Model

As in the previous chapter, we consider an isolated economic geographical system in which people carry out social and economic activities. Production is similar to that in the one-sector neoclassical growth model. The population is homogenous. It is assumed that product can be either invested or consumed. We assume that the total labor force is fully employed. Industrial good is selected to serve as numeraire. Households achieve the same utility level regardless of where they locate. Markets are perfectly competitive. The system is geographically linear and consists of two parts - the CBD and the residential area. The isolated state consists of a finite strip of land extending from a fixed central business district (CBD) with constant unit width (e.g., Solow and Vickrey, 1971, Suh, 1988). We assume that all economic activities are concentrated in the CBD. The residential area is occupied by households. We assume that the CBD is located at the left-side end of the linear territory. We introduce

N — the fixed population;
$K(t)$ — the total capital stocks at time t;
$w(t)$ and $r(t)$ — the wage rate and the rate of interest rate, respectively; and

7.1 The Model

$F(t)$ and $C(t)$ — the output of the production sector and the total consumption of the commodity, respectively.

The production is carried out by combination of capital and labor force in the form of

$$F = K^{\alpha}(zK)^{\beta}, \quad \alpha + \beta = 1, \quad \alpha, \beta > 0 \tag{7.1.1}$$

where z is human capital index. The marginal conditions are given by

$$r = \frac{\alpha F}{K}, \quad w = \frac{\beta F}{N}. \tag{7.1.2}$$

We now describe locational behavior of households. First, we introduce

L — the fixed (territory) length of the isolated state;
ω — the distance from the CBD to a point in the residential area;
$R(\omega,t)$ — the land rent at location ω at time t;
$k(\omega,t)$ and $s(\omega,t)$ — the capital stocks owned by and the savings made by a household at location ω, respectively;
$c(\omega,t)$ and $y(\omega,t)$ — the consumption and the net income of a household at location ω, respectively; and
$y(\omega,t)$ and $L_h(\omega,t)$ — the residential density and the lot size of a household at location ω.

According to the definitions of L_h and n, we have

$$n(\omega,t) = \frac{1}{L_h(\omega,t)}, \quad 0 \le \omega \le L. \tag{7.1.3}$$

It is assumed that each worker owns L/N amount of land and it is impossible to sell land but it is free to rent one's own land to others. This is analytically equal to the assumption of the public ownership in the literature, which means that the revenue from land is equally shared among the population. The total land revenue is given by

$$\overline{S}(t) = \int_0^L R(\omega,t)\, d\omega. \tag{7.1.4}$$

The income $\overline{s}$ from land per household is given by $\overline{s}(t) = \overline{S}(t)/N$. The net income $y(\omega,t)$ per household at location ω consists of three parts, the wage income, the income from land ownership and the interest payment for the household's capital stocks. The net income is thus given by

$$y(\omega,t) = r(t)y(\omega,t) + w(t) + \bar{s}(t). \tag{7.1.5}$$

We assume that utility level U of the household at location ω is dependent on the temporary consumption level $c(\omega,t)$, lot size $L_h(\omega,t)$, the leisure time $T_h(\omega,t)$, and the household's net wealth $k(\omega,t) + s(\omega,t) - \delta_k k(\omega,t)$, where δ_k ($1 > \delta_k > 0$) is the fixed depreciation rate of capital, in the following way

$$U(\omega,t) = T_h^\sigma c^\xi L_h^\eta (k + s - \delta_k k)^\lambda, \quad \sigma(t), \xi(t), \eta(t), \lambda(t) > 0,$$
$$\sigma(t) + \xi(t) + \eta(t) + \lambda(t) = 1 \tag{7.1.6}$$

where $\sigma(t), \xi(t), \eta(t)$, and $\lambda(t)$ are respectively the propensities to use leisure time, to consume goods, to consume housing, and to hold wealth. We use lot size to measure housing conditions and neglect possible complexity of housing production (e.g., Arnott, 1987, Anas, 1982, Bruecker, 1981, Bruecker and Rabenau, 1981, Hockman and Pines, 1980). Moreover, for convenience of representation, we assume $\delta_k = 0$. From the analysis below, it can be seen that this will not affect our essential conclusions.

By preference change we mean changes in $\sigma(t), \xi(t), \eta(t)$, and $\lambda(t)$. The form of utility functions in (7.1.6) is used in studies on economic growth with endogenous saving by Zhang (e.g., 1999, 2000). The above utility implies that at each point of time the consumer has a preference described in the form of utility function over his current consumption and wealth. In this model, a consumer determines two variables, how much he consumes and how much he saves when setting aside part of the current income $y(t)$ into the 'saving account'. It is easy to explain c and L_h in U. To explain $k + s - \delta_k k$, we may consider a situation in which the consumer can change his savings $p_k(t)k(t)$, with $p_k(t)$ ($=1$) being the price of capital stocks at t, into 'money' at any point of time without any 'transaction cost' or time delay. In other words, the consumer perceives $p_k(t)k(t)$ in the same way as he can treat his salary income at each point of time. In fact, we may perceive that at each pay-day the consumer 'mixes' his current income y and k and then decides how much he would spend on current consumption and how much he would put the total money in the saving account. In economics nothing should be free. Physical capital is subject to its laws of natural or social depreciation. Some one, for instance, the owner of physical capital, has to pay the depreciation. We assume that the consumer who owns the capital loses a fixed ratio δ_k of his past savings due to depreciation of physical capital. Hence, the consumer makes decision on $k + s - \delta_k k$ at each point of time. Obviously, we take account of the consumer's attitudes towards the future by the propensities to consume and to hold wealth.

In our approach we don't need a concept of the sum of utility over a period of time. In (7.1.6), we assume that an increase in the net wealth tends to increase utility level of a typical household. The net wealth may be accumulated for different reasons such as the capitalist spirit, old age consumption, providing education for children, power and social status (see e.g., Modigliani, 1986, Ram, 1982, Gersovitz, 1988, Cole, Mailath and Postlewaite, 1992, Fershtman and Weiss, 1993). Those different reasons determine preference structures. People's attitude towards the future in economic developing processes at each point of time is reflected in the (relative) value of the propensity $\lambda(t)$ to hold wealth.

It is necessary to remark that in a traditional intertemporal framework, the economy maximizes

$$\int_0^\infty U(C) \exp(-\rho t) dt$$

subject to the dynamic budget constraint of capital accumulation. In the above formula (Ramsey, 1928), there is a strict assumption. That is, utility is additional over time. It is not reasonable to add happiness over time. How to psychologically calculate expected future happiness and how to weigh the expected future against the present in temporary decision are both practically and theoretically difficult issues. To assume the existence of a discounting rate of human happiness over infinite time and to assume that people behave in such a way that they maximize the discounted utility are not only unreasonable, but also of little fruit from so many efforts of theoretical economists in the literature of economic dynamics. The parameter ρ in the above problem is meaningless if utility is not additional over time. It should be also remarked that in the traditional studies of growth and preference change (e.g., Uzawa, 1968, Wan, 1970, Boyer, 1978, Shi and Epstein, 1993, Auer, 1998), by preference change it means changes of ρ. Since ρ is defined with respect to utility, it is conceptually not easy to introduce reasonable dynamics of ρ. From the literature of economic growth, we know that it is quite difficult to discuss issues related to preference change and economic structure within the traditional optimal framework. It is obvious that our formula does not involve the issues just mentioned and can endogenously determine savings of the households. Our utility is not static. Happiness is flow and is determined as a function of consumption and net wealth at each point of time. It is wealth and consumption that are additional, rather than happiness, over time.

For simplicity, we specify $T_h(\omega)$ as follows

$$T_h(\omega) = T_0 - \upsilon\omega, \quad T_0 > 0. \tag{7.1.7}$$

The function $T_h(\omega)$ means that the leisure time is equal to the total available time T_0 minus the traveling time $\upsilon\omega$ from the CBD to the dwelling site. Here, we neglect possible impact of congestion and other factors on the traveling time (e.g., Arnott, 1979, Karlqvist, Lundqvist and Snickars, 1975, Anas, 1982, Roy and Johansson, 1993). For simplicity, we neglect traveling cost. We assume that the traveling time and locational housing conditions are the main factors that affect residential location. Although it is conceptually not difficult to take account of traveling cost in the budget constraint, it will cause analytical difficulties.

As the population is homogeneous and we neglect any transaction costs associated with residential changes, the utility level that people obtain should be equal over space at each point of time. We thus have

$$U(\omega_1, t) = U(\omega_2, t), \quad 0 \le \omega_1, \omega_2 \le L. \tag{7.1.8}$$

The income is distributed between consuming goods, housing and saving. The budget constraint is given by

$$c(\omega, t) + R(\omega, t)L_h(\omega, t) + s(\omega, t) = y(\omega, t) \tag{7.1.9}$$

at any point of time. Maximizing $U(\omega, t)$ subject to the budget constraint yields

$$c(\omega) = \xi\rho y(\omega) + \xi\rho k(\omega), \quad R(\omega)L_h(\omega) = \eta\rho y(\omega) + \eta\rho k(\omega)$$
$$s(\omega) = \lambda\rho y(\omega) - (\xi + \eta)\rho k(\omega) \tag{7.1.10}$$

where $\rho \equiv 1/(\xi + \eta + \lambda)$. The above equations mean that the housing consumption and consumption of the good are positively proportional to the net income and capital wealth, and the savings is positively proportional to the net income but negatively proportional to the wealth.

According to the definition of $s(\omega, t)$, we have the following capital accumulation for the household at location ω

$$\frac{dk(\omega)}{dt} = s(\omega), \quad 0 \le \omega \le L.$$

Substituting $s(\omega)$ in (7.1.10) into the above equation yields

$$\frac{dk(\omega)}{dt} = \lambda \rho y(\omega) - \delta k(\omega), \quad 0 \leq \omega \leq L \tag{7.1.11}$$

where $\delta \equiv (\xi + \eta)\rho \; (\lambda > \delta > 0)$.

As the economic system is isolated, the total population is distributed over the whole urban area. The population constraint is given by

$$\int_0^L n(\omega) d\omega = N. \tag{7.1.12}$$

The consumption constraint is given by

$$\int_0^L n(\omega) c(\omega) d\omega = C. \tag{7.1.13}$$

We also have

$$S(t) + C(t) = F(t) \tag{7.1.14}$$

where $S(t)$ is the total investment (= the total savings) of the society

$$S(t) = \int_0^L s(\omega,t) n(\omega,t) d\omega. \tag{7.1.15}$$

By the definitions of k, n and K, we have

$$\int_0^L n(\omega) k(\omega) d\omega = K. \tag{7.1.16}$$

From the definition of $y(\omega,t)$ in (7.1.5) and capital accumulation equation (7.1.11) we see that $k(\omega,t)$ and $y(\omega,t)$ are not explicitly dependent on location ω. That is, the capital stocks owned by per household and the net income are identical over space at any point of time. Multiplying (7.1.11) by $n(\omega,t)$ and then integrating the resulted equation from 0 to L with respect to ω yields

124   7 Urban Growth and Pattern Formation with Preference Change

$$\frac{dK}{dt} = \lambda \rho Y(\omega) - \delta K \qquad (7.1.17)$$

where $Y(t) \equiv y(t)N$ is the total income of the population at time.

We now specify possible dynamics of $\sigma(t)$, $\xi(t)$, $\eta(t)$, and $\lambda(t)$. As $\sigma(t) + \xi(t) + \eta(t) + \lambda(t) = 1$ holds at any point of time, we have to specify dynamics of three variables of the four. In order to further reduce dimension of preference dynamics, we assume that there exist relationships, $\sigma(t) = f_\sigma[\xi(t), \lambda(t)]$ and $\eta(t) = f_\eta[\xi(t), \lambda(t)]$ between σ, ξ, η, and λ at any point of time. For simplicity, we specify f_σ and f_η as follows

$$\sigma = \sigma_0 - h_\sigma \xi, \quad \eta = \eta_0 - h_\eta \xi \qquad (7.1.18)$$

where σ_0 and η_0 ($1 > \sigma_0, \eta_0 > 0$) are parameters. The two equations mean that the propensity $\sigma(t)$ to use leisure time and the propensity $\eta(t)$ to consume housing have linear relationships with the propensity $\xi(t)$ to consume commodities at any point of time. In general, it is difficult to explicitly provide signs of h_σ and h_η. It is possible that when the propensity $\xi(t)$ to consume commodities is increased, the propensity $\eta(t)$ to consume housing may either be increased or decreased, depending on the household's preference. This implies that the parameter h_η may be either positive or negative. Similarly, h_σ may be either positive or negative. By $\sigma + \xi + \eta + \lambda = 1$ and (7.1.18), we solve ξ as function of λ as follows

$$\xi = \xi_0 - h_\lambda \lambda \qquad (7.1.19)$$

where $\xi_0 \equiv (1 - \sigma_0 - \eta_0)h_\lambda$ and $h_\lambda \equiv 1/(1 - h_\sigma - h_\eta)$. We require $h_\lambda > 0$ and $1 > \xi_0 > 0$. This implies that an increase in the propensity λ to hold wealth is associated with a decrease in the propensity ξ to consume commodities. We have $h_\lambda > 0$ and $1 > \xi_0 > 0$ if $1 > \sigma_0 + \eta_0$ and $\sigma_0 + \eta_0 > h_\sigma + h_\eta$ which are assumed to be satisfied in the remainder of this chapter.

As $\sigma(t)$, $\xi(t)$ and $\eta(t)$ can be represented as functions of $\lambda(t)$ at any point of time, in order to model dynamics of preference it is sufficient for us to be concerned only with dynamics of $\lambda(t)$. Although we may generally argue that the propensity to hold wealth is affected by wealth and living conditions, it is not easy to generalize meaningful functional form of the preference change. In fact, many factors, such as demographic variables, income levels and wealth (e.g., Modigliani, 1986, Ram, 1982, Gersovitz, 1988), may have interactions with the propensity to own wealth. For simplicity, we assume that the propensity $\lambda(t)$ to hold wealth is affected by the capital stock $K(t)$ and the net income $Y(t)$ as follows

$$\frac{d\lambda}{dt} = \Phi(k, y) \equiv \theta\left(\frac{\theta_1}{1 + \theta_2 k^a y^b} - \lambda\right), \quad \lambda \geq 0, \quad \infty > \theta \geq 0$$

(7.1.20)

where θ, θ_1 (< 1) and θ_2 are positive parameters. If $\theta = 0$, λ is constant. This case means that preference is not affected by current living conditions and wealth accumulated. If $\theta \to +\infty$, we have

$$\frac{\theta_1}{1 + \theta_2 k^a y^b} = \lambda$$

at any point of time (except some singular points, e.g., Chow and Hale, 1982, Haken, 1983, O'Malley, 1988, Kevorkian and Cole, 1981). This implies that preference is quickly adapted to living conditions and wealth accumulated. If $a > (<) 0$, then the propensity to hold wealth tends to be reduced (increased) as the capital stock, k, per capita is increased. If $a = 0$, the term Φ is not affected by k. If $b > (<) 0$, then the propensity to hold wealth tends to be reduced (increased) as the income y per capita is increased. If $b = 0$, the term Φ is not affected by y. To illustrate the point that the parameters, a and b, may be either positive or negative, we like to quote from Fisher on habit formation (1930, pp.337-8)

> It has been noted that a person's rate of preference for present over future income, given a certain income stream, will be high or low according to the past habits of the individual. If he has been accustomed to simple and inexpensive ways, he finds it fairly easy to save and ultimately accumulate a little property. The habits of thrift being transmitted to the next generation, by imitation or by heredity or both, result in still further accumulation. The foundations of some of the world's greatest fortunes have been based upon thrift.
>
> If a man has been brought up in the lap of luxury, he will have a keener desire for present enjoyment than if he had been accustomed to the simple living of the poor. The children of the rich, who have been accustomed to luxurious living and who have inherited only a

fraction of their parent's means, may spend beyond their means and thus start the process of the dissipation of their family fortune. In the next generation this retrograde movement is likely to gather headway and to continue until, with the gradual subdivision of the fortune and the reluctance of the successive generations to curtail their expenses, the third or fourth generation may come to actual poverty.

The accumulation and dissipation of wealth do sometimes occur in cycles. Thrift, ability, industry and good fortune enable a few individuals to rise to wealth from the ranks of the poor. A few thousand dollars accumulation under favorable circumstances may grow to several millions in the next generation or two. Then the unfavorable effects of luxury begin, and the cycle of poverty and wealth begins anew. The old adage, 'From shirt sleeves to shirt sleeves in four generations,' has some basis in fact.

When analyzing saving behavior of national economies, we may similarly argue that the parameters a and b are dependent on the stage of economic development. We will examine what will happen to the dynamics when a and b are taken on different values. We require that for $\lambda > 0$, $1 > \sigma$, ξ, $\eta > 0$. It can be seen that the requirement is satisfied under appropriate constraints on values of the parameters, σ_0, η_0, η_σ and η_λ.

We have thus built the growth model with endogenous spatial distribution of wealth, consumption and the population, capital accumulation and preference change. The system has 21 variables, k, c, L_h, s, n, T_h, U, R, y, F, K, C, w, r, S, $\overline{S}$, and $\overline{s}$, σ, ξ, η, and λ. It contains the same number of equations. We now show that the problem has solutions.

7.2 Properties of the Dynamic System

Before examining the dynamic properties of the system, we show that the dynamics can be described by the motion of two variable, K and λ.

First, multiplying the equations in (7.1.10) by $n(\omega,t)$ and then integrating these equations from 0 to L with respect to ω, we obtain

$$T_h = \eta\rho(Y + K), \quad C = \xi\rho(Y + K), \quad S = \lambda\rho Y - (\xi + \eta)\rho K \tag{7.2.1}$$

where we use (7.1.4), (7.1.13) and (7.1.15). Substituting S and C in (7.2.1) into (7.1.14) yields

7.2 Properties of the Dynamic System

$$Y = \frac{F + \eta\rho K}{(\xi + \lambda)\rho}. \tag{7.2.2}$$

As ξ, η, and ρ are functions of λ and F is a function of K, we conclude that the net income Y is a function of K and λ at any point of time. By (7.1.17) and (7.1.20), we have

$$\frac{dK}{dt} = \lambda\rho Y - \delta K, \quad \frac{d\lambda}{dt} = \theta\left(\frac{\theta_1}{1 + \theta_2 K^a Y^b / N^{a+b}} - \lambda\right). \tag{7.2.3}$$

The two-dimensional dynamic system (7.2.3) has two variables K and λ. We can show that all the other variables are uniquely determined as a function of K, λ and ω ($0 \leq \omega \leq L$) at any point of time. The following lemma is proved in the appendix.

Lemma 7.2.1.
The dynamics of $K(t)$ and $\lambda(t)$ are determined by (7.2.3). For any given $K(t)$ (> 0) and $\lambda(t)$ ($\theta_1 > \lambda > 0$), at any point of time, all the other variables in the system are uniquely determined as functions of $K(t)$, $\lambda(t)$ and ω ($0 \leq \omega \leq L$).

We thus explicitly determine the motion of the system over time and space. We now guarantee existence of equilibria. Equilibrium is given as a solution of the following equations

$$\lambda\rho Y = \delta K, \quad \frac{\theta_1}{1 + \theta_2 K^a Y^b / N^{a+b}} = \lambda. \tag{7.2.4}$$

By $\lambda\rho Y = \delta K$ in (7.2.4) and (7.2.2), we solve

$$K = \left(\frac{\lambda}{\xi}\right)^{1/\beta} zN. \tag{7.2.5}$$

Substituting $\lambda\rho Y = \delta K$ and (7.2.5) into the second equation in (7.2.4), we have

$$H(\lambda) \equiv \frac{\theta_2 z^{a+b} \lambda^x (\xi + \eta)^b}{\xi^{(a+b)/\beta}} + \lambda - \theta_1 = 0 \tag{7.2.6}$$

where $x \equiv (\beta + a + \alpha b)/\beta$. Since a and b may be either positive or negative, we see that x may be either positive or negative. In the remainder of this chapter, we require $x > 0$, i.e., $\beta + a + \alpha b > 0$. In the case of $a > 0$ and $b > 0$, this is always held. When a and/or b are negative, this requirement (i.e., $\beta > -a - \alpha b$) means that households' propensity to hold wealth is not highly increased when the capital and income are increased. If $x < 0$, we see that the system may have multiple equilibria.

Since $x > 0$, we have $H(0) < 0$ and $H(\theta_1) > 0$. This guarantees that $H(\lambda) = 0$ has at least one solution for $0 < \lambda < \theta_1$. Hence, the equilibrium problem has at least one solution. Taking derivatives of (7.2.6) with respect to λ yields

$$\frac{dH}{d\lambda} = 1 + \frac{h_\lambda \theta_2 z^{a+b} \lambda^x (1 + \eta/\xi)}{\beta \xi^{(a+b)/\beta}} \left[\frac{\beta x \xi}{h_\lambda \lambda} + \frac{b(\alpha + \beta h_\eta + \eta/\xi)}{1 + \eta/\xi} + a \right] \quad (7.2.7)$$

in which (7.1.18) and (7.1.19) are used. We notice that $h_\lambda \geq 0$ and

$$\beta h_\eta + \frac{\eta}{\xi} = -\alpha h_\eta + \frac{\eta_0}{\xi} > 0.$$

In the case of $a > 0$ and $b > 0$, we have $H' > 0$. Since $H(0) < 0$, $H(\theta_1) > 0$, $H' > 0$, we conclude that $H(\lambda) = 0$ has a unique solution for $0 < \lambda < \theta_1$. By (7.2.5) and (7.2.6), we uniquely determine λ and K. The other variables are uniquely determined by Lemma 7.2.1. It can be seen that if

$$\frac{b(\alpha + \beta h_\eta + \eta/\xi)}{1 + \eta/\xi} + a \geq 0$$

then the system has a unique equilibrium. For instance, in the case of $b < 0$, $a > 0$, and $a > -b$, this inequality is held. But in general, it is difficult to judge whether the system has a unique equilibrium when a and/or b are negative.

By (7.2.3), (7.2.2) and the equilibrium conditions, we calculate the two eigenvalues, ϕ_1 and ϕ_2, as follows

$$\phi^2 + \left[\frac{\beta\lambda F}{(\xi + \lambda)K} + b_2\right]\phi + \frac{\beta b_2 \lambda F / K + (F + h_\lambda K)b_1}{\xi + \lambda} = 0$$

(7.2.8)

where

$$b_1 = \frac{\theta^*}{K}\left[a + \frac{(\alpha F / \rho K + \eta)b\lambda}{(\xi + \lambda)^2}\right],$$

$$b_2 \equiv \frac{\theta\lambda}{\theta_1} + \theta^*\left[\frac{1-b}{\lambda} + \frac{b\xi}{\lambda(\xi + \lambda)} + \frac{(\lambda h_\eta + \eta_0)b\eta_\lambda}{\eta(\xi + \lambda)}\right],$$

$$\theta^* \equiv \frac{\theta\theta_2\lambda^2 K^a Y^b}{\theta_1 N^{a+b}} > 0.$$

(7.2.9)

The system is stable if

$$\frac{\beta\lambda F}{(\xi + \lambda)K} + b_2 > 0, \quad \beta b_2\lambda F/K + (F + h_\lambda K)b_1 > 0.$$

Obviously, in the case of $a > 0$, $1 \geq b \geq 0$, $h_\eta \leq \eta_0$, the system is stable. Summering the above discussion, we get the following proposition.

Proposition 7.2.1.
We assume $x > 0$. The system has at least one equilibrium. In the case of $a > 0$, $1 \geq b \geq 0$, the dynamic system has a stable unique equilibrium.

The assumption of $a > 0$ and $b > 0$ is important for guaranteeing the existence of a unique equilibrium. If we relax this assumption, the system may have multiple equilibria. This means that the dynamic system is stable if the consumers decrease their propensity to save when they hold more wealth; otherwise it is possible that the system becomes unstable if consumers tend to save too much when they become more wealthy.

7.3 Human Capital and the Economic Structure

In order to show interdependence between capital accumulation, preference, residential location, wages and incomes over space, we are now concerned with effects of change in the level of human capital on the equilibrium economic structure.

For illustration, we require $a > 0$ and $b > 0$. Hence, the system has a stable equilibrium. First, taking derivatives of (7.2.6) with respect to z yields

$$H' \equiv \frac{(a+b)(\lambda - \theta_1)}{z} < 0 \qquad (7.3.1)$$

where $H' > 0$ is defined by (7.2.7) and $\lambda - \theta_1 > 0$ by (7.2.6). An increase in the level z of human capital reduces the propensity λ to hold wealth. As people work more effectively, their propensity to save is reduced. By (7.1.18) and (7.1.19), we have

$$\frac{d\sigma}{dz} = h_\sigma h_\lambda \frac{d\lambda}{dz}, \quad \frac{d\eta}{dz} = h_\eta h_\lambda \frac{d\lambda}{dz}, \quad \frac{d\xi}{dz} = -h_\lambda \frac{d\lambda}{dz} > 0. \qquad (7.3.2)$$

The propensity ξ to consume goods is increased. The sign of $d\sigma/dz$ ($d\eta/dz$) is the opposite to that of h_σ (h_η).

By (7.2.5) and $\lambda \rho Y = \delta K$, we get

$$\frac{\beta}{K}\frac{dK}{dz} = \frac{\beta}{z} + \left(\frac{1}{\lambda} + \frac{h_\lambda}{\xi}\right)\frac{d\lambda}{dz},$$

$$\frac{\beta}{Y}\frac{dY}{dz} = \frac{\beta}{z} + \left[\frac{\alpha}{\lambda} + \frac{h_\lambda}{\xi} + \frac{(h_\eta - 1)\beta h_\lambda}{\xi + \eta}\right]\frac{d\lambda}{dz}. \qquad (7.3.3)$$

As $d\lambda/dz$ is negative, the signs of dK/dz and dY/dz are ambiguous. If the propensity to hold wealth is almost not affected by changes in human capital (i.e., $d\lambda/dz$ being negligible), then the capital K and net income Y are increased as human capital is improved. When the propensity to hold wealth is strongly reduced as the human capital is improved, then the level of capital stocks and the net income may be reduced. We see that when we explicitly take account of preference change, the impact of improved human capital on wealth and income are more complicated than the traditional neoclassical growth theory which holds that an improvement in human capital always brings about higher equilibrium levels of the capital stocks and incomes. We see that this is not necessarily held in our model. It should be remarked that our analysis is conducted under presumed $a > 0$ and $b > 0$. If we relax this requirement, it is more difficult to explicitly judge the impact of change in z.

Taking derivatives of (7.1.1) with respect to z yields

7.3 Human Capital and the Economic Structure

$$\frac{1}{F}\frac{dF}{dz} = \frac{1}{z} + \left(\frac{1}{\lambda} + \frac{h_\lambda}{\xi}\right)\frac{\alpha}{\beta}\frac{d\lambda}{dz}. \tag{7.3.4}$$

The sign of dF/dz is ambiguous. By (7.1.2), we have

$$\frac{dw}{dz} = \frac{w}{F}\frac{dF}{dz}, \quad \frac{dr}{dz} = -\left(\frac{1}{\lambda} + \frac{h_\lambda}{\xi}\right)r\frac{d\lambda}{dz} > 0. \tag{7.3.5}$$

The impact on the wage rate has the same sign as that on the output level. The rate of interest is increased as human capital is improved. By (7.2.1) and $Y = (\xi + \eta)K/\lambda$, we have

$$\frac{1}{C}\frac{dC}{dz} = \frac{1}{z} + \left(\frac{1}{\lambda} + \frac{h_\lambda}{\xi}\right)\frac{\alpha}{\beta}\frac{d\lambda}{dz},$$

$$\frac{1}{\overline{S}}\frac{d\overline{S}}{dz} = \frac{1}{z} + \left(\frac{h_\eta h_\lambda}{\eta} + \frac{\alpha}{\beta\lambda} + \frac{h_\lambda}{\beta\xi}\right)\frac{d\lambda}{dz}. \tag{7.3.6}$$

If $d\lambda/dz$ is negligible, then the consumption level C and land revenue $\overline{S}$ are increased as the level of human capital is increased. By (7.A.1.3) and (7.A.1.2), we have

$$\frac{1}{n(0)}\frac{dn(0)}{dz} = h_\sigma\left(h_\lambda + \frac{\sigma}{\eta}\right)$$

$$\left[\frac{1}{\eta + \sigma} + T_0(1 - \upsilon L/T_0)^{1+\sigma/\eta}\frac{\ln(1 - \upsilon L/T_0)}{\eta \upsilon L}\right]\frac{d\lambda}{dz},$$

$$\frac{1}{n(\omega)}\frac{dn(\omega)}{dz} = \frac{1}{n(0)}\frac{dn(0)}{dz} + h_\sigma\left(h_\lambda + \frac{\sigma}{\eta}\right)\frac{\ln(1 - \upsilon L/T_0)}{\eta}\frac{d\lambda}{dz}. \tag{7.3.7}$$

In (7.3.7), the term $\ln(1 - \upsilon L/T_0)$ is negative. It is not easy to explicitly interpret the impact on the spatial distribution of the population. A special case is $h_\sigma = 0$. This case simply means that the households' propensity to use leisure time is constant. Change in economic conditions and propensities to hold wealth or to consume goods have no impact on the propensity to use leisure time. In this case, we have: $dn(\omega)/dz = 0$. Hence, the residential distribution is not affected by improvement in human capital. The urban pattern formation in terms of the residential distribution is constant over space even though living conditions in terms of consumption and wealth are improved as human capital is improved. But when h_σ is not equal to zero, then the residential distribution over space is affected by change in z. If we don't further specify values of some parameters, it is difficult to explicitly judge the sign of $dn(\omega)/dz$ for any $0 \leq \omega \leq L$.

By (7.2.10), we have

$$\frac{1}{R(\omega)}\frac{dR(\omega)}{dz} = \frac{1}{n(\omega)}\frac{dn(\omega)}{dz} + \left(\frac{h_\eta h_\lambda}{\eta} + \frac{\alpha}{\beta\lambda} + \frac{h_\lambda}{\beta\xi}\right)\frac{d\lambda}{dz} + \frac{1}{z}.$$

(7.3.8)

Since $dn(\omega)/dz$ may be either positive or negative, $d\lambda/dz$ is negative, and $1/z$ is positive, we see that it is difficult to explicitly judge the impact on land rent. Since the last two terms in the right-hand side are independent of ω, we see that $dR(\omega)/dz$ tends to be positive (negative) when $dn(\omega)/dz$ is positive (negative) at location ω. That is, the land rent tends to increase (decrease) at the location where the residential density is increased (decreased). We see that the land rent distribution is affected by human capital in a complicated way. It is interesting to notice that when $dn(\omega)/dz = 0$ in the case of $h_\sigma = 0$, $dR(\omega)/dz$ is not necessarily equal to zero. In other words, even when an improvement in human capital has no impact on the residential distribution pattern, the land rent is affected by the change.

Since we explicitly solved the equilibrium, we may also analyze the impact of changes in any other variable on the economic geography.

7.4 On Preference Change and Urban Pattern

This paper proposed an economic model with urban pattern formation. The model describes a dynamic interdependence between economic growth, preference change, residential location, wages and incomes over space. The production and capital accumulation are modeled within the framework of neoclassical growth theory. Urban location is analyzed within the framework of the neoclassical urban economics. We introduced endogenous preference change by assuming that the propensities are dependent on living conditions and wealth accumulated by the population. Conditions for existence of equilibria and stability are provided. The effects of changes in human capital on the economic structure were examined.

The synthesis provides some insights into complex of interactions between different economic forces. It can be seen that we can be certain about the uniqueness and stability of economic geography only under conditions that the propensity to hold wealth is weakened when the wealth and income are increased. It is known that urban economists tend to find population and knowledge dynamics as sources of instabilities and multiple equilibria. This chapter points out another possibility of instabilities and multiple equilibria in evolution of economic geography. If the propensity to hold wealth is strengthened when the wealth and income are increased, then the economic geography may have multiple equilibria and instabilities. Although geographical aspects of our model are oversimplified in comparison to contemporary literature of economic geography, our model shows how the residential distribution pattern and land rent distribution may be related to economic growth. We show this relationship by analyzing the impact of changes in human capital on the economic geography. We may extend the model in this chapter in different ways. For instance, it is quite important to introduce endogenous dynamics of human capital in the long-run study of spatial economic evolution. As preference change is so complicated, it is quite difficult to propose dynamics of preference, which may be called 'general'. It is necessary to carry out empirical examination in order to find out proper patterns of preference changes.

Appendix

A.7.1 Proving Lemma 7.2.1

We determine σ, ξ, η, Y and F as functions of K and λ. First, we note that $k = K/N$ and $y = Y/N$ hold. By (7.1.2), w and r are determined. By $\overline{S} = \eta \rho (Y + K)$ in (7.2.1) and $\bar{s} = \overline{S}/N$, we have $\overline{S}$ and $\bar{s}$. By (7.1.10), $S = sN$ and $C = cN$, we determine c, s, C and S are determined.

Substituting $L_h = 1/n$ and $T_h(\omega) = T_0 - \upsilon\omega$ into (7.1.6), we have

$$U(\omega) = c^\xi (T_0 - \upsilon\omega)^\sigma \frac{(k+s)^\lambda}{n^\eta(\omega)}. \tag{7.A.1.1}$$

Using $U(\omega) = U(0)$, we have

$$n(\omega) = n(0)\left(1 - \frac{\upsilon\omega}{T_0}\right)^{\sigma/\eta} \tag{7.A.1.2}$$

Substituting (7.A.1.2) into (7.1.12) yields

$$n(0) = \frac{\upsilon N}{T_0}\left(1 + \frac{\sigma}{\eta}\right)\left[1 - \left(1 - \frac{\upsilon\omega}{T_0}\right)^{1+\sigma/\eta}\right]. \tag{7.A.1.3}$$

By (7.A.1.2) and (7.A.1.3), we determine $n(\omega)$. By $L_h = 1/n$ and the first equation in (7.1.10), we directly get L_h and R. We thus completed the proof of the lemma.

8 Urban-Rural Division of Labor with Spatial Amenities

This chapter proposes a spatial model within which urban and rural areas are assumed to have different locational characteristics. The model is influenced by the studies related to location of firms with exogenous spatial distribution of demand (e.g., Isard, 1956, Greenhut, 1963, Greenhut, Norman and Hung, 1989) and the studies related to economic geography of residential area with endogenous income and production structure (e.g., Alonso, 1964, Beckmann, 1957, Wingo, 1961, Muth, 1969, Stull, 1974, David and Rosenbloom, 1990, Fujita, 1989, Calem and Carlino, 1991). This chapter constructs a model that bears on problems inherent in economic geography with interactions of demand and supply of various commodities, price structure and spatial division of labor with given preference, territory, labor force and locational amenities. The remainder of the chapter is organized as follows. Section 8.1 represents the basic model. Section 8.2 provides conditions for existence of equilibria. Section 8.3 examines effects of changes in the urban amenity on spatial economic structure. Section 8.4 concludes the study. The appendix proves the results in Section 8.2. This chapter is based on Zhang (1997a).

8.1 The Spatial Structure with Urban-Rural Areas

This chapter features a linear system on a homogeneous plain whose width is unity. For simplicity, the CBD is assumed to be one-side edged as shown in Fig.8.1.1. The urban area is located between the CBD and the farm. It is assumed that farmers' dwelling and farming sites have the same location. Possible costs for professional changes and transportation costs for commodities are neglected. The labor markets are assumed to be perfectly competitive. The population is homogenous and people are freely mobile between the two professions. The agricultural good is selected to serve as numeraire, with all other prices being measured relative to its price.

To describe the system, we define

L and N — the given territorial size and the population of the system;
L_1 — the boundary between the urban area and the farm;
ω — the residential location of a worker, $0 \leq \omega \leq L_1$;

8 Urban-Rural Division of Labor with Spatial Amenities

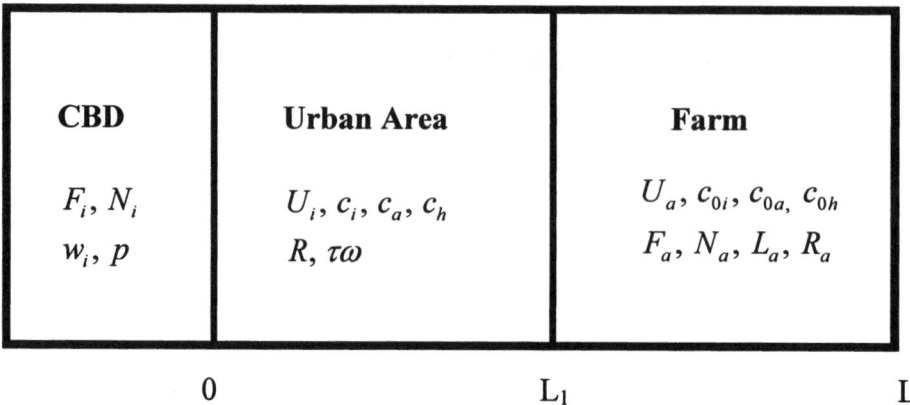

Fig. 8.1.1. The One-City and One-Farm System

$R(\omega)$ — the land rent at location ω in the urban area;

R_a — the land rent of the agricultural area;

N_i and N_a — the labor employed by the industrial and agricultural sectors, respectively;

F_i and F_a — the output of the industrial and agricultural sectors, respectively;

p — the price of the urban product; and

w_i and w_a — the wage rates of the workers and farmers, respectively.

The agricultural sector
There are two, labor and land, inputs in agricultural production. The agricultural production function is specified as follows

$$F_a = L_a^\alpha N_a^{1-\alpha}, \quad 0 < \alpha < 1 \tag{8.1.1}$$

where L_a is land employed by the agricultural sector. The production function is homogenous of degree one with respect to its two inputs.

The profit maximization yields the following conditions

$$R_a = \frac{\alpha F_a}{L_a}, \quad w_a = \frac{(1-\alpha)F_a}{N_a}. \tag{8.1.2}$$

8.1 The Spatial Structure with Urban-Rural Areas

Let C_i and C_a denote total consumption of agricultural product by the workers and the farmers, respectively. The balance of demand for and supply of agricultural product is given by

$$C_i + C_a = F_a. \qquad (8.1.3)$$

The industrial sector
It is assumed that industrial commodity is produced by a single input factor - labor. Other possible inputs such as natural resources and capital are omitted at this initial stage. A linear production function of the industrial sector is specified as follows

$$F_i = zN_i. \qquad (8.1.4)$$

The parameter z measures productivity of the industrial sector.

Perfect competition in the labor and land markets results in zero profit of the industrial sector. That is, $pF_i = w_i N_i$, or

$$zp = w_i. \qquad (8.1.5)$$

The balance of demand for and supply of industrial commodity is given by

$$C_{ii} + C_{ia} = F_i \qquad (8.1.6)$$

where C_{ii} and C_{ia} are the total consumption levels of industrial commodity by the workers and farmers, respectively.

The behavior of consumers
It is assumed that the utility level of a typical household is dependent on locational amenity and consumption levels of industrial commodity, agricultural goods, housing. Here, a household's housing is measured simply by its dwelling size. The utility function of an urban household at location ω is specified as follows

$$U_i(\omega) = Ac_i^\xi(\omega)c_a^\mu(\omega)L_h^\eta(\omega), \quad \xi, \mu, \eta > 0, \quad 0 \le \omega \le L_1 \qquad (8.1.7)$$

where $c_i(\omega)$, $c_a(\omega)$, and $L_h(\omega)$ are respectively consumption levels of industrial commodity, agricultural good and dwelling size of the household at location ω, and ξ, μ, and η are parameters.

Similarly, the utility function of a typical farmer is specified as follows

$$U_a = A_a c_{ai}^{\xi} c_{aa}^{\mu} L_{ah}^{\eta}, \quad \xi, \mu, \eta > 0 \tag{8.1.8}$$

where c_{ai}, c_{aa}, and L_{ah} are, respectively, consumption levels of industrial commodity, agricultural good and dwelling size of each farmer.

The two parameters, A_i and A_a, are called amenity levels of city and countryside, respectively. As life styles, and 'professional amenities' may be different between the city and the countryside, it is reasonable to assume that even if the components of consumption are identical between a worker and a farmer, they may have different utility levels. The parameters, A_i and A_a, take account of difference in amenities due to professional, locational and other factors. Although amenity levels are treated as parameters in this chapter, it is not difficult to endogenously determine A_i and A_a as functions of population and production scales in the ways as suggested, for instance, in Kanemoto (1980).

The urban consumer problem is defined by

$$\text{Max } U_i(\omega), \quad \text{s.t.: } pc_i(\omega) + c_a(\omega) + R(\omega)L_h(\omega) = y(\omega) \tag{8.1.9}$$

where

$$y(\omega) = w - \upsilon\omega, \quad 0 \le \omega \le L_1.$$

Here, the parameter υ is travel cost per unity of distance and $\upsilon\omega$ is the total traveling cost per household between the dwelling site ω and the CBD.

The optimal problem has the following unique solution

$$c_i(\omega) = \frac{\xi \rho y(\omega)}{p}, \quad c_a(\omega) = \mu \rho y(\omega), \quad L_h(\omega) = \frac{\eta \rho y(\omega)}{R(\omega)} \tag{8.1.10}$$

where $\rho \equiv 1/(\xi + \mu + \eta)$.

The rural consumer problem is formulated by

$$\text{Max } U_a(\omega), \quad \text{s.t.: } pc_{ai} + c_{aa} + R_a L_{ah} = w_a.$$

The unique optimal solution is given by

8.1 The Spatial Structure with Urban-Rural Areas

$$c_{ai} = \frac{\xi \rho w_a}{p}, \quad c_{aa} = \mu \rho w_a, \quad L_{ah}(\omega) = \frac{\eta \rho w_a}{R_a}. \tag{8.1.11}$$

Denoting $n(\omega)$ the residential density at location ω, we have

$$n(\omega) = \frac{1}{L_h(\omega)}, \quad 0 \le \omega \le L_1. \tag{8.1.12}$$

As the number of the workers is equal to the sum of the urban residents over the urban area and the total population consists of workers and farmers, the following equations are held

$$\int_0^{L_1} n(\omega) d\omega = N_i, \quad N_i + N_a = N. \tag{8.1.13}$$

The total land of the agricultural area is utilized for farmers' housing and agricultural production, i.e.

$$L_a + L_{ah} N_a = L - L_1. \tag{8.1.14}$$

As the system is isolated, agricultural and industrial product are consumed by all the households in the system. The conditions, (8.1.3) and (8.1.6), can be expressed in the following forms

$$\int_0^{L_1} c_i(\omega) n(\omega) d\omega + (L - L_1) c_{ai} = F_i,$$

$$\int_0^{L_1} c_a(\omega) n(\omega) d\omega + (L - L_1) c_{aa} = F_a. \tag{8.1.15}$$

The model has thus been completed. It has 20 variables, N_i, N_a, F_i, F_a, c_i, c_a, L_h, c_{ai}, c_{aa}, L_{ah}, w_i, w_a, p, L_1, L_a, R_a, $R(\omega)$, $n(\omega)$, U_i, and U_a. It contains the same number of equations.

8.2 Spatial Equilibria

As the population is homogenous, utility level is identical among the households, irrespective of dwelling location and professions. Using (8.1.7), (8.1.10) and $U_i(\omega) = U_i(0)$ for $0 \leq \omega \leq L_1$, we have

$$\frac{R(\omega)}{R(0)} = \left[\frac{y(\omega)}{w_i}\right]^\lambda < 1, \quad 0 \leq \omega < L_1 \tag{8.2.1}$$

where $y(\omega) \leq w_i$ and $\lambda \equiv 1/\eta\rho > 1$. It is direct to see that urban area's land rent declines as it is further from its CBD. As the urban households have identical utility function and wage rate, the conclusion is obvious.

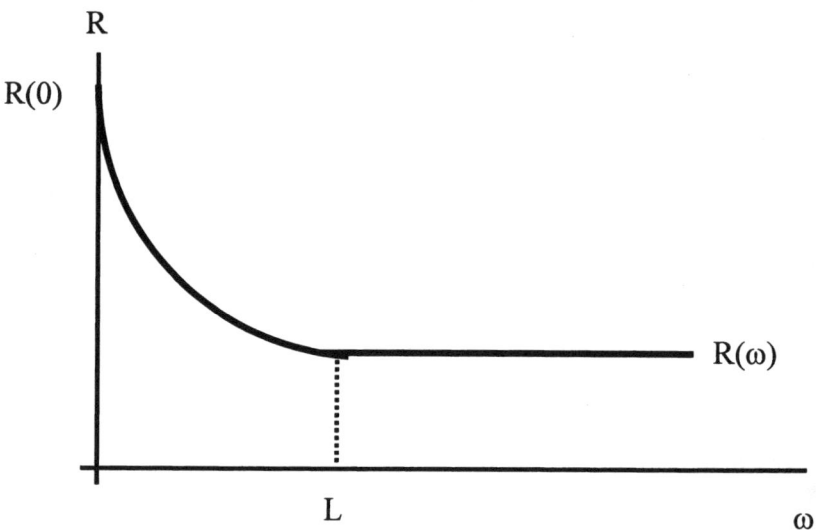

Fig.8.2.1. Rent as Functions of the Distance

Substituting (8.1.7), (8.1.8), (8.1.10) and (8.1.11) into $U_i(L_1) = U_a$ and $U_i(0) = U_a$, respectively, yield

$$\frac{y(L_1)}{w_a} = A^{1/\lambda}, \quad \frac{R(0)}{R_a} = A\left(\frac{w_i}{w_a}\right)^\lambda \tag{8.2.2}$$

where $A \equiv (A_i / A_a)^{\lambda\rho}$. If the rural amenity level is much lower than that in the city, farmers' wage rate may be higher than workers. This is due to the assumed homogenous population. The following proposition shows that the system has always equilibria.

Proposition 8.2.1.
The wage rate w_i of workers is determined by the following equation

$$\Phi^*(w_i) \equiv \left[\frac{1}{Z(w_i)} - 1\right]vw_i - [vL - w_i + w_i Z(w_i)]A_i = 0 \quad (8.2.3)$$

where $Z(w_i)$ is a continuous function of w_i, defined in Appendix A.3.1. The equation, (8.2.3), has at least one positive solution. Moreover, for any solution, w_i (> 0), all other variables are uniquely determined as functions of w_i.

The above proposition is proved in the appendix. Here, the existence of equilibria is guaranteed. But the system may have either a unique or multiple equilibria. It should be remarked that for any solution w_i of (8.2.3), the equilibrium values of all other variables are uniquely determined as functions of w_i.

8.3 Amenities and Economic Geography

This section examines impact of changes in the urban amenity level A_i on production structure and the economic geography. Taking derivatives of (8.A.1.14) and (8.A.1.17) with respect to A_i yields

$$\frac{1}{w_i}\frac{dw_i}{dA_i} + a_1 \frac{dZ}{dA_i} = a_2,$$

$$\frac{v\theta_1 L}{w_i^2}\frac{dw_i}{dA_i} - \left(\frac{\lambda}{Z^{1+\lambda}} + \theta_1\right)\frac{dZ}{dA_i} = \frac{2Z^\lambda}{(1-Z^\lambda)\eta A_i} > 0 \quad (8.3.1)$$

in which Z, θ_1 and θ_2 are defined in Appendix A.3.1 and

$$a_1 \equiv \frac{\alpha\lambda}{(1-\alpha)(1-Z^\lambda)Z} + \frac{\xi Z + \mu A^{1/\lambda}}{[\xi Z + (1-\alpha)\mu A^{1/\lambda}]Z} > 0,$$

8 Urban-Rural Division of Labor with Spatial Amenities

$$a_2 \equiv \frac{2\alpha/\eta + \rho - \alpha\rho}{1-\alpha} + \frac{\alpha\xi\mu A^{1/\lambda}}{\left[\xi Z + (1-\alpha)\mu A^{1/\lambda}\right]A_i} > 0. \tag{8.3.2}$$

From (8.3.1) we get

$$\left(\frac{\lambda}{Z^{1+\lambda}} + \theta_1 + \frac{\upsilon\theta_1 a_1 L}{w_i}\right)\frac{dZ}{dA_i} = a_3,$$

$$\frac{\upsilon\theta_1 L}{w_i^2}\frac{dw_i}{dA_i} = \left(\frac{\lambda}{Z^{1+\lambda}} + \theta_1\right)\frac{dZ}{dA_i} + \frac{2Z^\lambda}{(1-Z^\lambda)\eta A_i} \tag{8.3.3}$$

where

$$a_3 \equiv \frac{a_2 \upsilon \theta_1 L}{w_i} - \frac{2Z^\lambda}{(1-Z^\lambda)\eta A_i}.$$

If $a_3 > 0$, $dZ/dA_i > 0$ and $dw_i/dA_i > 0$. As $Z = 1 - \upsilon L_1/w_i$, in the case of $a_3 > 0$ we have $d(\upsilon L_1/w_i)/dA_i < 0$. This implies that the travel cost of urban residents from the CBD to the boundary of the urban area is reduced in the term of city's wage rate. If $a_3 < 0$, then $dZ/dA_i < 0$ (i.e., $d(\upsilon L_1/w_i)/dA_i > 0$), but dw_i/dA_i may be either positive or negative. As a_3 is determined by so many parameters it is not easy to explicitly interpret the sign of a_3.

Taking derivatives of $L_1 = (1-Z)w_i/\upsilon$ with respect to A_i yields

$$\frac{w_i}{L_1}\frac{dL_1}{dA_i} = \frac{dw_i}{dA_i} - \frac{w_i}{1-Z}\frac{dZ}{dA_i}. \tag{8.3.4}$$

We conclude that dL_1/dA_i may be either positive or negative. It is obviously not easy to interpret the condition (8.3.4). From $p = w_i/z$ we see that dp/dA_i has the same sign as that of dw_i/dA_i. When city's amenity level is increased, the price of industrial commodity will be expanded in the case that city's wage rate is increased.

Taking derivative of (8.A.1.10) with respect to A_i yields

8.3 Amenities and Economic Geography

$$\frac{dL_a}{dA_i} = -\frac{\alpha_0}{\alpha_0 + 1/\lambda}\frac{dL_1}{dA_i}. \tag{8.3.5}$$

If the urban area is expanded due to improvement in city's amenity level, the land employed by the agricultural sector will be reduced, and vice versa.

From

$$w_a = \frac{y(L_1)}{A^{1/\lambda}}, \quad N_a = \left(\frac{1-\alpha}{w_a}\right)^{1/\alpha}, \quad N_i = N - N_a$$

we obtain

$$A^{1/\lambda}\frac{dw_a}{dA_i} = \frac{dw_i}{dA_i} + \left(1 + \frac{1}{\lambda\alpha_0}\right)v\frac{dL_a}{dA_i} - \frac{py(L_1)}{A_i},$$

$$\frac{1}{N_a}\frac{dN_a}{dA_i} = -\frac{1}{\alpha w_a}\frac{dw_a}{dA_i} + \frac{1}{L_a}\frac{dL_a}{dA_i},$$

$$\frac{dN_i}{dA_i} = -\frac{dN_a}{dA_i}. \tag{8.3.6}$$

The sign of dw_a/dA_i is jointly determined by three terms, dw_i/dA_i, $(1 + 1/\lambda\alpha_0)vdL_a/dA_i$, and $-py(L_1)/A_i$. One may call them, respectively, effects of changes in city's amenity level on city's wage rate, the land employed by the agricultural sector and rural attractiveness. If an improvement in city's amenity level results in increases of city's wage rate and agricultural land use, the wage rate of the agricultural sector is increased. If the wage rate of the agricultural sector is increased and the land use of the agricultural sector is reduced, the employment of the agricultural sector is reduced.

Taking derivatives of $R_a = \alpha(N_a/L_a)^{1-\alpha}$ with respect to A_i, we obtain

$$\frac{1}{(1-\alpha)R_a}\frac{dR_a}{dA_i} = \frac{1}{N_a}\frac{dN_a}{dA_i} - \frac{1}{L_a}\frac{dL_a}{dA_i}. \tag{8.3.7}$$

Substituting the second equation in (8.3.6) into (8.3.7) yields

$$\frac{dR_a}{dA_i} = -\frac{(1-\alpha)}{\alpha w_a} R_a \frac{dw_a}{dA_i}. \tag{8.3.8}$$

The sign of dR_a/dA_i is the opposite to that of dw_a/dA_i. From $R(0) = (w_i/w_a)^\lambda AR_a$, the first equation in (8.3.6), and (8.3.8) one gets

$$\frac{1}{R(0)}\frac{dR(0)}{dA_i} = \frac{1}{\eta A_i} + \frac{\lambda}{w_i}\frac{dw_i}{dA_i} - \left(\frac{1-\alpha}{\alpha w_a} + \frac{\lambda}{w_a}\right)\frac{dw_a}{dA_i}. \tag{8.3.9}$$

If the wage rates are not much affected by changes in city's amenity, i.e., dw_i/dA_i and dw_a/dA_i being small, the land rent at the CBD is increased as the city's amenity is improved. Taking derivatives of $R(\omega) = R(0)(1 - \upsilon\omega/w_i)^\lambda$ with respect to A_i yields

$$\frac{1}{\lambda R(\omega)}\frac{dR(\omega)}{dA_i} = \frac{\rho}{A_i} + \left(\frac{\upsilon\omega}{w_i - \upsilon\omega} + \frac{1}{w_i}\right)\frac{dw_i}{dA_i},$$

$$-\left(\frac{1-\alpha}{\alpha\lambda w_a} + \frac{1}{w_a}\right)\frac{dw_a}{dA_i}, \quad 0 < \omega < L_1. \tag{8.3.10}$$

If city's wage rate is increased but rural wage rate is reduced due to improvement in city's amenity, then land rent at any location in the urban area will be increased. But if dw_a/dA_i is positive, the impact on the urban rent may be either positive or negative.

From (8.1.10), (8.1.11) and (8.1.12) and the above results, we get the impact of changes in A_i on $n(\omega)$, and the consumption components of the farmers and urban workers.

8.4 On Spatial Equilibrium Structure

This chapter developed a one-city and one-farm spatial equilibrium model with endogenous spatial pattern in a linear system. We emphasized the impact of amenity difference between cities and division of labor on economic geography. The system is described by the interactions among the 20 variables of spatial division of labor, urban and rural outputs, wages, land use and land rent distribution and consumption components of the farmers and urban workers over space. We showed how these variables can be explicitly expressed as functions of the parameters in the system. We also examined the effects of changes in city's amenity level on the variables.

The model may be extended in different ways. For instance, it is necessary to introduce certain mechanisms to include amenities and productivity as endogenous variables in a long-run dynamic analysis. It is not difficult to introduce endogenous technology and capital accumulation. There are many factors for scale and scope economies which are important for explaining urban centralization and decentralization. The significance of introducing scale and scope economies is essentially important when dealing with economic systems consisting of multiple regions and nations. It is also necessary to relax some strict assumptions on the geographical structure designed in this chapter.

Appendix

A.8.1 Proving Proposition 8.2.1.

First, by (8.1.10), (8.1.13) and (8.2.1) we have

$$n(\omega) = \frac{1}{L_h(\omega)} = \left[\frac{y(\omega)}{w_i}\right]^\lambda \frac{\lambda R(0)}{y(\omega)}. \tag{8.A.1.1}$$

Substituting (8.A.1.1) into (8.1.14) yields

$$1 - \left[\frac{y(L_1)}{w_i}\right]^\lambda = \frac{\upsilon N_i}{R(0)}. \tag{8.A.1.2}$$

Substituting (8.A.1.1), (8.1.11) and (8.1.12) into (8.1.16) yields

$$G = \frac{pF_i}{\xi} = \frac{F_a}{\mu} \tag{8.A.1.3}$$

where

$$G \equiv \frac{\left[w_i^{1+\lambda} - y^{1+\lambda}(L_1)\right]\lambda\rho R(0)}{\upsilon w_i^\lambda (1+\lambda)} + \rho w_a(L - L_1). \tag{8.A.1.4}$$

From (8.1.2) we have $L_a = \alpha_0 w_a N_a / R_a$, where $\alpha_0 \equiv \alpha/(1-\alpha)$. Substituting this equation and $L_{ah} = w_a / \lambda R_a$ into (8.1.14) yields

$$\left(\alpha_0 + \frac{1}{\lambda}\right)\frac{w_a N_a}{R_a} = L - L_1. \tag{8.A.1.5}$$

From (8.A.1.3) we obtain: $F_a = \mu p F_i / \xi$. From this equation and $w_a = (1-\alpha) F_a / N_a$, the following equation is held

$$w_a = \frac{(1-\alpha)\mu p F_a}{\xi N_a}. \tag{8.A.1.6}$$

Substituting (8.2.2) and $zp = w_i$ into (8.A.1.6) yields

$$\frac{N_i}{N_a} = \frac{\xi y(L_1)}{(1-\alpha)\mu A^{1/\lambda} w_i}. \tag{8.A.1.7}$$

From (8.2.2), (8.A.1.2) and (8.A.1.5), the following equation is held

$$\frac{N_i}{N_a} = \left[\frac{w_i}{y(L_1)}\right]^\lambda \left[1 - \left\{\frac{y(L_1)}{w_i}\right\}^\lambda\right]\frac{(\alpha_0 + 1/\lambda)y(L_1)A^{2-1/\lambda}}{\upsilon(L-L_1)}. \tag{8.A.1.8}$$

From (8.A.1.7) and (8.A.1.8) we get

$$\left[\frac{w_i}{y(L_1)}\right]^\lambda \left[1 - \left\{\frac{y(L_1)}{w_i}\right\}^\lambda\right] = \frac{\theta_i(L-L_1)}{w_i} \tag{8.A.1.9}$$

where

$$\theta_i = \frac{\lambda \upsilon \xi}{(\lambda \alpha_0 + 1)(1-\alpha)\mu A^2}.$$

From $w_a N_a / R_a = L_a / \alpha_0$ and (8.A.1.5) we obtain

$$L_a = \frac{(L - L_1)\lambda\alpha_0}{\lambda\alpha_0 + 1}. \tag{8.A.1.10}$$

From $w_a = (1 - \alpha)(L_a / N_a)^\alpha$, (8.A.1.10) and $y(L_1)/w_a = A^{1/\lambda}$, we have

$$N_a = \frac{(1 - \alpha)^{1/\alpha}(L - L_1)\alpha_0 A^{1/\alpha\lambda}}{(\alpha_0 + 1/\lambda)y^{1/\lambda}(L_1)}. \tag{8.A.1.11}$$

From $N_i = N - N_a$ and (8.A.1.7), N_a is given by

$$N_a = \frac{(1 - \alpha)\mu w_i A^{1/\lambda} N}{(1 - \alpha)\mu w_i A^{1/\lambda} + \xi y(L_1)}. \tag{8.A.1.12}$$

From (8.A.1.11) and (8.A.1.12) we have

$$L - L_1 = \frac{\theta_2 w_i y^{1/\alpha}(L_1)}{(1 - \alpha)\mu w_i A^{1/\lambda} + \xi y(L_1)} \tag{8.A.1.13}$$

in which

$$\theta_2 \equiv \frac{(\alpha_0 + 1/\lambda)(1 - \alpha)\mu A^{1/\lambda} N}{\alpha_0[(1 - \alpha)A^{1/\lambda}]^{1/\alpha}} > 0.$$

The two equations, (8.A.1.9) and (8.A.1.13), contain two variables, w_i and L_1. It is now shown that the two equations have solutions. Substituting $(L - L_1)$ in (8.A.1.9) into (8.A.1.13), the following equation is held

$$\Phi(w_i, Z) \equiv w_i^{1-1/\alpha}(1 - Z^\lambda)[\xi Z + (1 - \alpha)\mu A^{1/\lambda}] - \theta_1\theta_2 Z^{\lambda+1/\alpha} = 0 \tag{8.A.1.14}$$

where $Z \equiv y(L_1)/w_i$. We now show that for any positive w_i there is a unique Z, $0 < Z < 1$ such that $\Phi(w_i, Z) = 0$. As

$$\Phi(w_i, 0) = (1 - \alpha)\mu A^{1/\lambda} w_1^{1-1/\alpha} > 0, \quad \Phi(w_i, 1) = -\theta_1\theta_2 < 0. \tag{8.A.1.15}$$

8 Urban-Rural Division of Labor with Spatial Amenities

There is, at least, one Z, $0 < Z < 1$, such that $\Phi(w_i, 0) = 0$. As the second partial derivative of Φ with respect to Z is negative for $w_i > 0$ and $0 < Z < 1$, the solution is unique. Let $Z(w_i)$ denote the unique solution of $\Phi(w_i, Z) = 0$. It should be remarked that it is easy to check that $dZ(w_i)/dw_i$ may be either positive or negative, depending on the parameter values. As $Z(w_i) = y(L_1)/w_i$, L_1 can be expressed as a unique function of w_i as follows

$$L_1(w_i) = \frac{[1 - Z(w_i)]w_i}{\upsilon}. \tag{8.A.1.16}$$

Substituting this relation into (8.A.1.9) yields (8.2.3).

It is now shown that the equation $\Phi^*(w_i) = 0$, has solutions. It is easy to check that $\Phi^*(0) < 0$ and $\Phi^*(w_i) > 0$ for sufficiently large w_i are held. This implies that $\Phi^*(w_i) = 0$ has positive solutions. Although it is not easy to find out under what conditions $\Phi^*(w_i) = 0$ has a unique solution, it is now shown that for any positive w_i the other 19 variables can be determined as unique functions of w_i. By (8.A.1.16) L_1 is given as a unique function of w_i. The price of industrial commodity p is given by $p = w_i / z$. From (8.A.1.10) one gets L_a as a function of L_1. From (8.A.1.11) and $N_i = N - N_a$, the labor distribution, L_i and N_a, is determined as functions of w_i and L_1. The outputs F_i and F_a are given by $F_i = zN_i$, $F_a = L_a^\alpha N_a^{1-\alpha}$. The land rent of the rural area and wage rate in the agricultural sector R_a and w_a are given by $R_a = \alpha F_a / L_a$ and $w_a = (1 - \alpha) F_a / N_a$, respectively. From (8.1.11) we obtain c_{ai}, c_{aa}, and L_{ah}, as functions of w_a, R_a and p. We get U_a and U_i. The land rent $R(0)$ at the CBD is given by $R(0) = (w_i / w_a)^\lambda A R_a$. The land rent $R(\omega)$ at any location in the urban area is determined by

$$R(\omega) = \left(1 - \frac{\upsilon \omega /}{w_i}\right)^\lambda R(0), \quad 0 \leq \omega < L_1.$$

From (8.1.10) one can determine the consumption components, $c_i(\omega)$, $c_a(\omega)$, and $L_h(\omega)$, of an urban household at location ω as functions of p, w_i, w and $R(\omega)$. The residential density $n(\omega)$ at location ω is directly given by $n(\omega) = 1/L_h(\omega)$. It has thus been shown that the 19 variables can be explicitly expressed as functions of w_i.

The proof of Proposition 8.2.1 was completed.

9 Spatial Equilibrium with Multiple Cities

Time and space are the two most essential factors for explaining economic reality. Any economic activity takes place at certain place at certain time. Although economic dynamics have caused much attention, the complexity of economic geography has largely been ignored in the mainstream of theoretical economics. It has now become clear that there are a number of potentially important spatial influences, such as public goods, amenities, different externalities, transportation costs, that may challenge the validity of competitive equilibrium theory for explaining a regionally heterogeneous economy. For instance, one of these factors is the so-called capitalization, which implies that the price of land is interdependent with local amenities, economic agents' densities, transportation costs and other local variables or parameters. Although the significance of capitalization has been noticed by location theorists (Scotchmer and Thisse, 1992), it may be said that we still have no compact framework within which we can satisfactorily explain the issue. Extending the model of the previous chapter, this chapter attempts to suggest a framework to model an economic geographical equilibrium with capitalization.

The seminar paper on compensating regional variation in wages and rents by Roback (1982) has caused a wide interest among regional and urban economists to theoretically investigate how the value of location attributes is capitalized into wages and services. Since the publication of Roback's model, many empirical and theoretical studies have shown that between urban areas wages may capitalize differences in amenity levels or living costs (e.g., Blomquist, Berger and Hoehn, 1988, Simon and Love, 1990, Bell, 1991, Voith, 1991). But as far as I know, only a few urban models with endogenous residential structure in the literature both take division of labor into account and endogenously determine incomes of households within a perfectly competitive framework.

In this chapter, we classify production into different sectors. In particular, we emphasize the geographical character of services in our modeling. Services are consumed simultaneously as they are produced and thus cannot be transported like commodities. Accordingly, when explicitly modeling economic geography, we have to take account of this special character of services. Many services such as schools, hospitals and restaurants have to be consumed where they are supplied. Accordingly, services have special location property in comparison to commodity production. When dealing with economic geography, we have to explain how spatial parameters or slow changing variables, such as infrastructures, city culture and climates, may affect attractiveness of the location under consideration. For simplicity of discussion, we measure these various factors, in an aggregated term, by a single variable of urban amenity (e.g., Diamond and Tolley, 1981). It is obvious that some location amenities such as pollution level, residential density and transportation congestion are dependent upon economic agents' activities, while other amenities such as climates, transportation structure, and historical buildings, may not be strongly affected by economic agents' activities, at least, within a short-run period of time. Accordingly, in a strict sense, it is necessary to classify amenities into endogenous and exogenous amenities when modeling economic geography. Which kinds of amenities should be classified as endogenous or exogenous also depend upon time scale of the analysis and the economic system under consideration.

This chapter is organized as follows. Section 9.1 defines the two-city and three-sector equilibrium model with endogenous residential structure. Section 9.2 provides conditions for existence of equilibria. Section 9.3 examines impact of changes in city 1's amenity upon the urban structure. Section 9.4 concludes the chapter. In the appendix we show how we can explicitly solve all the variables in terms of the parameters. This chapter is based on Zhang (1993d).

9.1 The Model

We consider an economic system consisting of two cities, indexed by 1 and 2, respectively. We assume that the spatial pattern of each city is similar to that of the standard urban economic model suggested by Alonso (1964). The basic issue of the Alonso model is to determine urban equilibrium pattern with the assumption that all the households maximize utility levels subject to the exogenously given incomes. Utility levels are dependent upon consumption levels of a composite good and land size for housing. Urban economists have made many efforts to generalize and extend the Alonso model (e.g., Straszheim, 1987, Sasaki, 1990). Unfortunately, in almost all of those models, incomes are exogenously given, even though a few models both with endogenous income and with residential structure have recently been suggested within a compact framework (e.g., Rauch, 1991).

Similarly to the Alonso model, we assume that each city consists of two parts - the CBD and a residential area; and the locations of the CBDs are pre-specified points and all product activities are concentrated at the CBDs. We feature a linear two-city system on a homogeneous plain whose width is a unit distance (Fig. 9.1.1). In economic geography, different spatial patterns have been discussed (e.g., Beckmann and Puu, 1985, Greenhut, Norman, and Hung, 1987). It has become clear that geographical patterns may make economic analysis extremely complicated. To get some explicit conclusions from our framework with endogenous residential location and incomes, we accept this almost simplest economic geographical arrangement with multiple city centers.

152   9   Spatial Equilibrium with Multiple Cities

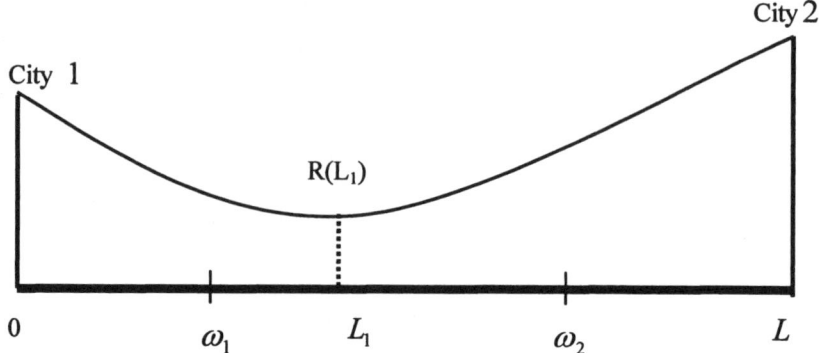

Fig.9.1.1. The Spatial Structure of the Two-City System

We assume that the system produces two kinds of consumption commodities, indexed by 1 and 2, respectively. We assume that production is characterized by urban specification and that commodity j is produced at CBD j. We neglect transportation costs of commodities. The CBDs are not only centers of economic production but also places for services such as school, hospitals, hotels and restaurants. Services are city-specified in the sense that services are consumed only by the households of the city which supply the services. Here, we neglect the possibility that services supplied in one city are consumed by the households from the other (such as tourism). Accordingly, each city has two - commodity and services — production sectors. As we neglect transportation cost of goods, any commodity is sold at the same price in the two cities. But services may have different prices in the two cities as they cannot be transported between the cities.

We assume that labor markets are characterized by perfect competition and that people are freely mobile. For simplicity of analysis, we assume that commodities are produced by only one input factor — labor. We neglect other inputs such as capital and land. We assume a homogeneous labor force and select commodity 1 to serve as numeraire, with all the other prices being measured relative to its price. Although they may be different between the two cities, the wage rates for different sectors in the same city are identical. The assumption of the homogeneous labor force and labor mobility results in identical utility level in the entire system. We introduce

L — fixed distance between the CBDs;

L_1 — distance from the CBD 1 to the boundary of the two residential areas;

ω_j — dwelling location of city j's residents, $0 \leq \omega_1 \leq L_1$ and $L_1 \leq \omega_2 \leq L$;

$R(\omega_j)$ — land rent at location ω_j, $j = 1, 2$;

N_{ij} and N_{sj} — labor force employed by industrial sector and service sector in city j, respectively;

N_j — city j's employment, $N_j = N_{ij} + N_{sj}$;

N — the total labor force of the system, $N = N_1 + N_2$;

F_{ij} and F_{sj} — product of the industrial sector and service sector of city j, respectively;

p — price of commodity 2;

p_{sj} — prices of services in city j;

w_j — wage rate in city j.

We now describe the model.

Supply of commodities and services
We specify a linear production function of industrial sector j as follows

$$F_{i1} = z_1 N_{i1}, \quad F_{i2} = z_2 N_{i2} \tag{9.1.1}$$

where z_j is city j's production efficiency parameter. As there is only one input factor in each production, we always have $F_{i1} = w_1 N_{i1}$ and $pF_{i2} = w_2 N_{i2}$, i.e.

$$w_1 = z_1, \quad w_2 = pz_2. \tag{9.1.2}$$

City 1's wage rate is equal to industry 1's product value per labor input; while city 2's wage rate is equal to industry 2's product value per labor input. As the labor markets are perfectly competitive and production functions are linear with a single input, we see that (9.1.2) provide two 'accounting relations' of the industries.

Let C_{kj} denote total consumption of city k's industrial product by city j. Then, the balances of demand for and supply of the two commodities are given by

$$C_{1j} + N_{2j} = F_{ij}, \quad j = 1, 2. \tag{9.1.3}$$

We assume that services are produced by a single input — labor. We specify the production function of the service sectors as follows

$$F_{sj} = zN_{sj}, \quad j = 1, 2 \tag{9.1.4}$$

where z is production efficiency parameter of the service sectors. We assume that service production has identical efficiency between the two cities. We neglect possible difference in quality of service supplies between the two cities. As become clear later on, this assumption can be easily relaxed by introducing $F_{sj} = zN_{sj}$ without affecting our main conclusions.

As we assume that the labor markets are perfectly competitive and labor force is homogeneous, the wage rate of the service sector in city j is identical to that of the industrial sector in the same city. As labor is the only input in service production, we have $w_j N_{sj} = p_{sj} F_{sj}$, i.e.

$$w_j = z p_{sj}, \quad j = 1, 2. \tag{9.1.5}$$

As services are consumed simultaneously as they are produced we always have

$$C_{sj} = F_{sj}, \quad j = 1, 2 \tag{9.1.6}$$

where C_{sj} is consumption of services of city j.

Demand structure of households and residential structure
We assume that the utility level of an individual from consuming commodities, services and housing can be expressed in the form of

$$U(\omega_j) = A_j c_1^{\xi}(\omega_j) c_2^{\mu}(\omega_j) c_s^{\gamma}(\omega_j) L_h^{\eta}(\omega_j), \quad \xi, \upsilon, \gamma, \eta > 0 \tag{9.1.7}$$

where A_j is amenity level of city j, ξ, υ, γ, and η are parameters, and $c_1(\omega_j)$, $c_2(\omega_j)$, $c_s(\omega_j)$ and $L_h(\omega_j)$, $j = 1, 2$, are, respectively, consumption levels of commodity 1, commodity 2, services and housing of a household at location ω_j in city j.

This chapter explicitly takes difference of amenities between the two cities into account. Although we admit that there are interactions among amenities and residential and companies' locations, for convenience of discussion we assume that amenity levels in each city are given in this chapter. It should be remarked that this assumption may be relaxed in different ways. The difference in amenity levels between the two cities implies that even if the components of consumption are identical among the households in the two cities, they may have different utility levels.

The consumer problem is defined by

$$\max\ U(\omega_j)$$
$$\text{s.t.:}\ c_1(\omega_j) + pc_2(\omega_j) + p_{sj}c_s(\omega_j) + R(\omega_j)L_h(\omega_j) = y(\omega_j) \tag{9.1.8}$$

in which

$$y(\omega_1) \equiv w_1 - \upsilon\omega_1,\ 0 \leq \omega_1 \leq L_1,$$
$$y(\omega_2) \equiv w_2 - \upsilon(L - \omega_2),\ L_1 \leq \omega_2 \leq L.$$

Here, υ is the travel cost per unity of distance and $\upsilon\omega_j$ is the total traveling cost between dwelling site ω_j and the CBD j.

The consumer problems has unique optimal solution as follows

$$c_1(\omega_j) = \xi\rho y(\omega_j),\ c_2(\omega_j) = \frac{\mu\rho y(\omega_j)}{p},\ c_s(\omega_j) = \frac{\gamma\rho y(\omega_j)}{p_{sj}},$$
$$R(\omega_j)L_h(\omega_j) = \eta\rho y(\omega_j),\ j = 1, 2 \tag{9.1.9}$$

in which $\rho \equiv 1/(\xi + \mu + \gamma + \eta)$. Here, we assume that w_j, L_0 and υ are taken on appropriate values so that any point in the interval $[0, L_0]$ is occupied by some resident.

We denote $n(\omega_j)$ the residential density at dwelling site ω_j. According to the definitions, we have

$$n(\omega_1) = \frac{1}{L_h(\omega_1)},\ 0 \leq \omega_1 \leq L_1,$$
$$n(\omega_2) = \frac{1}{L_h(\omega_2)},\ L_1 \leq \omega_2 \leq L. \tag{9.1.10}$$

We have the following constraints upon the population distribution between the two urban areas

$$\int_0^{L_1} n(\omega_1)d\omega_1 = N_1,\ \int_{L_1}^{L} n(\omega_2)d\omega_2 = N_2. \tag{9.1.11}$$

The constraints (9.1.3) and (9.1.6) can be expressed in the following forms

$$\int_0^{L_1} n(\omega_1) c_{1j}(\omega_1) d\omega_1 + \int_{L_1}^{L} n(\omega_2) c_{2j}(\omega_2) d\omega_2 = F_{ij}, \quad j = 1, 2,$$

$$\int_0^{L_1} n(\omega_1) c_s(\omega_1) d\omega_1 = F_{s1}, \quad \int_{L_1}^{L} n(\omega_2) c_s(\omega_2) d\omega_2 = F_{s2}. \quad (9.1.12)$$

We have thus completed constructing the equilibrium model. The system consists of 23 variables, N_{ij}, N_{sj}, N_j, F_{ij}, F_{sj}, w_j, p_{sj}, $(j = 1, 2)$, c_1, c_2, c_s, n, R, p, L, U and also of the same number of independent equations. We now show that these variables can be explicitly expressed as functions of the parameters, z_1, z_2, A_1, A_2, z, and L.

9.2 Urban Equilibria

This section shows that the 23 endogenous variables can be expressed as functions of the parameters in the system. As we neglect any cost for migration between the two cities and assume the identical preference among the households, utility level for any household must be identical in the interval $[0, L_0]$. Substituting (9.1.9) into $U(\omega_j) = U(L_1)$ yields

$$\frac{R(\omega_j)}{R(L_1)} = \left[\frac{y(\omega_j)}{y_j(L_1)}\right]^m > 1, \quad j = 1, 2 \quad (9.2.1)$$

where

$$m \equiv \frac{1}{\eta \rho}, \quad y_1(L_1) \equiv w_1 - vL_1, \quad y_2(L_1) = w_2 - v(L - L_1).$$

As $y(\omega_j)/y(L_1) > 1$, we see that $R(\omega_j)/R(L_1) > 1$. The equations state that land rent of any urban area declines as the residential location is further away from its CBD. As the workers employed in a city have the identical wage rate and traveling cost is positively proportional to the distance, we see that housing price will decline as the workers travel further from their dwelling sites to the working place.

9.2 Urban Equilibria

Lemma 9.2.1.
The land rent of an urban area declines as the distance from the CBD is increased. That is, $dR/d\omega_1 < 0$, for $0 \leq \omega_1 \leq L_1$ and $-dR/d\omega_2 < 0$, for $L_1 \leq \omega_2 \leq L$.

Substituting (9.1.9) into the relation of $U(L_1)$ for city 1 being equal to $U(L_1)$ for city 2 yields

$$\frac{p_{s1}}{p_{s2}} = \left(\frac{A_1}{A_2}\right)^{1/\gamma} \left[\frac{y_1(L_1)}{y_2(L_1)}\right]^{1/\gamma\rho}. \tag{9.2.2}$$

Service prices between the two cities are different when urban amenities and wages are different between the two cities. In the case of $y_1(L_1) = y_2(L_1)$, if city 1's amenity is higher that in city 2, city 1's service price is higher than that in city 2, and vice versa. As $y_1(L_1) = w_1 - \upsilon L_1$ and $y_2(L_1) = w_2 - \upsilon(L - L_1)$, if the service price in a city with a higher wage rate tends to be higher than that in the other city. In Fig.9.2.1, we illustrate two cases when $y_1(L_1) = y_2(L_1)$.

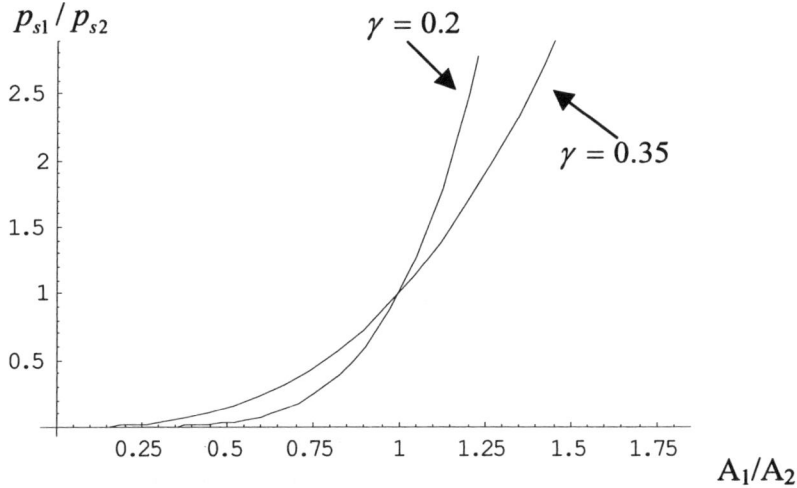

Fig.9.2.1. Relations Between Service Prices and Amenities

Substituting (9.1.9) into $U(0) = U(L_1)$ together with (9.2.2) yields

$$\frac{R(0)}{R(L)} = \left[\frac{w_1 y_2(L_1)}{w_2 y_1(L_1)}\right]^m. \tag{9.2.3}$$

The land rents between the residential areas at the CBDs will be different if $w_1 = w_2$ and $L_1 = L/2$ are not held.

The following proposition which provides conditions for existence of equilibria is proved in the appendix.

Proposition 9.2.1.
The equilibrium price p of commodity 2 is determined by

$$\Phi(p) = 0, \quad 0 < \Lambda(p) < L_1 \tag{9.2.4}$$

in which $\Phi(p)$ and $\Lambda(p)$ are continuous functions of p with $\Phi(p)$ defined in (9.A.1.11) and $\Lambda(p)$ in (9.2.6). Moreover, for any positive equilibrium price p, the other variables are uniquely determined as functions of p. In particular, the service prices and the boundary between the two cities are given by

$$p_{s1} = \frac{z_1}{z}, \quad p_{s2} = \frac{z_2 p}{z}, \tag{9.2.5}$$

$$L_1 = \frac{z_1 p^{\gamma \rho} - z_2 Ap + A\upsilon L}{(p^{\gamma \rho} + A)\upsilon} \equiv \Lambda(p) \tag{9.2.6}$$

in which

$$A \equiv \left[\left(\frac{A_2}{A_1}\right)\left(\frac{z_1}{z_2}\right)^\gamma\right]^\rho.$$

From (9.A.1.11) we see that $\Phi(p) = 0$ is a complicated function. We discuss under what conditions the equation may have solutions in the appendix. We will not examine the conditions in detail, as they are too difficult for explicit interpretations. The condition, $p_{s1} = z_1/z$, is derived from the condition of perfect competition in city 1's labor market. As labor is the only input in production and the production functions take on linear forms, the equality of the wage rate in city 1's two sectors guarantees the condition. We see that city 1's service price is constant (in the term of the city's industrial product) as z_1 and z are fixed parameters. We can similarly interpret $p_{s2}/p = z_2/z$. Equation (9.2.6) determines the boundary between the two cities as a function of the price of city 2's industrial product. In Appendix, we show how all the other variables can be uniquely determined as functions of p.

9.3 The Impact of Amenities upon the Urban Structure

This section examines the impact of changes in city 1's amenity A_1 upon the urban system. First, from (9.2.5) we have

$$\frac{dp_{s1}}{dA_1} = 0, \quad \frac{dp_{s2}}{dA_1} = \left(\frac{z_2}{z_1}\right)\frac{dp}{dA_1}. \tag{9.3.1}$$

Here, we can get dp/dA_1 directly by taking derivative of (9.2.4) with respect to A_1. The price of services in city 1 is not affected by changes in city 1's amenity. The impact of changes in city 1's amenity upon the price of services in city 2 has the same sign as that upon the price of commodity 2. Here, we will not explicitly represent dp/dA_1 as it is too complicated to interpret. The sign of A_1 may be either positive or negative, depending upon combination of different parameters.

Taking derivatives of (9.2.6) with respect to A_1 yields

$$\upsilon(p^{\gamma p} + A)\frac{dL_1}{dA_1} = \frac{\gamma p p^{\gamma p} y_1(L_1) - Aw_2}{p}\frac{dp}{dA_1} + \frac{w_2 - \upsilon L}{A_1}\rho A$$
$$+ \frac{A\gamma \rho L_1}{A_1} \tag{9.3.2}$$

in which we require $w_2 - \upsilon L_0 > 0$. If the price is not much affected, i.e., $|(1/p)dp/dA_1|$ being small, then $dL/da_1 > 0$. If $|(1/p)dp/dA_1|$ is small, so is $|(1/p_{s2})dp_{s2}/dA_1|$. As amenity in city 1 is increased, city 1 becomes more attractive. Since the prices are little affected, more people will move to city 1. Accordingly, city 1's urban area will be expanded. From (9.2.2) and (9.2.5) we have $p^{\gamma p} y_1(L_1) = Ay_2$. The term, $\gamma p p^{\gamma p} y_1(L_1) - Aw_2$, in (9.3.2) is equal to $A(\gamma p y_2(L_1) - w_2)$. As $\gamma p < 1$ and $\gamma p y_2(L_1) < w_2$, we see that

$\gamma p p^{\gamma p} y_1(L_1) - Aw_2 < 0$. We conclude that if $dp/dA_1 < 0$, dL_1/dA_1 is always positive. But if $dp/dA_1 > 0$, dL_1/dA_1 may be either positive or negative.

From (9.A.1.14) we have the following impact upon the land rent at the boundary

$$\frac{dR(L_1)}{dA_1} = \upsilon N \frac{dN_0}{dA_1} \qquad (9.3.3)$$

in which

$$\frac{1}{\upsilon m N_0^2} \frac{dN_0}{dA_1} = -\frac{z_1^m}{y_1^{m+1}(L_1)} \frac{dL_1}{dA_1}$$

$$+ \left(\frac{pz_2}{y_2(L_1)}\right)^m \frac{dL_1/dA_1 - (L_1/p)dp/dA_1}{y_2(L_1)}. \qquad (9.3.4)$$

We see that if $|(1/p)dp/dA_1|$ is small, then $dR(L_1)/dA_1 > 0$. Also, if $dp/dA_1 < 0$, $dR(L_1)/dA_1 > 0$. When the price of commodity 2 is reduced, the land rent at the boundary is increased. When $dp/dA_1 > 0$, the impact of changes in city 1's amenity upon the land rent at the boundary may be either positive or negative.

From (9.A.1.13), we have the following impact upon the population distribution between the two cities

$$\frac{dN_1}{dA_1} = \frac{N_1}{N_0} \frac{dN_0}{dA_1} + \frac{\upsilon m z_1^m N N_0}{y_1^{m+1}(L_1)} \frac{dL_1}{dA_1},$$

$$\frac{dN_2}{dA_1} = -\frac{dN_1}{dA_1}. \qquad (9.3.5)$$

We see that when $dp/dA_1 < 0$, city 1's employment will be increased but city 2's employment is reduced. If $dp/dA_1 > 0$, it is difficult to judge the sign of dN_1/dA_1.

From (9.A.1.9), (9.A.1.10), (9.A.1.15), (9.1.9) and the above analytical results we can directly provide the impact of changes in A_1 upon the variables,

$R(\omega_j)$, N_{ij}, N_{sj}, $c_1(\omega_j)$, $c_2(\omega_j)$, $c_s(\omega_j)$, and $L_h(\omega_j)$, $j = 1, 2$. We will not represent the results here.

We now compare how the consumption components of the households in the two cities are affected by changes in city 1's amenity. For simplicity, we just compare the consumption components of the households at the two CBDs. From $y(0) = w_1 = z_1$, $y(L) = w_2 = pz_2$ and (9.1.9) we have

$$\frac{c_1(0)}{c_2(L)} = \frac{z_1}{pz_2}, \quad \frac{c_2(0)}{c_2(L)} = \frac{z_1}{pz_2}, \quad \frac{c_s(0)}{c_s(L)} = 1. \tag{9.3.6}$$

Taking derivatives of (9.3.6) with respect to A_1 yields

$$\frac{d(c_1(0)/c_2(L))}{dA_1} = \frac{d(c_2(0)/c_2(L))}{dA_1} = -\frac{z_1}{p^2 z_2}\frac{dp}{dA_1} > 0,$$

$$\frac{d(c_s(0)/c_s(L))}{dA_1} = 0. \tag{9.3.7}$$

The ratio of service consumption per household between the households at the CBD 1 and at the CBD 2 is not affected by changes in city 1's amenity. If commodity 2's price is decreased as city 1's amenity is improved, $c_1(0)/c_2(L)$ and $c_2(0)/c_2(L)$ are increased, and vice versa. From the condition $U(0)/U(L) = 1$ and (9.3.7) we can directly have that if $dp/dA_1 < 0$, then $d(L_h(0)/L_h(L))/dA_1 > 0$.

From the above discussion we see that only when $dp/dA_1 < 0$, it is possible to explicitly judge the impact of changes in city 1's amenity upon the urban structure. When $dp/dA_1 > 0$, it is difficult to explicitly judge the impact without further specifying the parameter values. As city 1's amenity is improved but its relative wage is reduced, the conclusion is expectable.

9.4 Concluding Remarks

We proposed a two-center model with income and output levels being endogenously determined. The model showed that wages, service prices, consumption components and land rent are different between the two cities at a perfectly competitive equilibrium. We also examined the impact of changes in city 1's amenity level upon the economic geography. We solved the problem under some strict assumptions. For instance, the linear spatial pattern is too simple. We also omit important inputs such as capital and land in the production functions. It is obvious that we may extend the model by using more general utility functions.

Appendix

A.9.1 Proving Proposition 9.2.1

In this appendix, we show how to get each equation in Proposition 9.2.1 and solve all the other variables as functions of the parameters in the system. From (9.1.9) and $R(\omega_j) = R(L_1)[y(\omega_j)/y_j(L_1)]^m$, we have

$$n(\omega_j) = y_j^{m-1}(\omega_j) Y_j(L_1) \tag{9.A.1.1}$$

where $Y_j(L_1) \equiv mR(L_1)/y_j^m(L_1)$. Substituting (9.1.9) and (9.A.1.1) into (9.1.11) and (9.1.12) then integrating the formula yield

$$w_j^m - y_j^m(L_1) = \frac{\upsilon m N_j}{Y_j}, \quad j = 1, 2, \tag{9.A.1.2}$$

$$Y_1^* + Y_2^* = \frac{(m+1)F_{i1}}{\xi \rho}, \quad Y_1^* + Y_2^* = \frac{(m+1)pF_{i2}}{\mu \rho}, \tag{9.A.1.3}$$

$$Y_1^* = \frac{(m+1)p_{s1}F_{s1}}{\gamma \rho}, \quad Y_2^* = \frac{(m+1)p_{s2}F_{s2}}{\gamma \rho} \tag{9.A.1.4}$$

where

$$Y_j^* \equiv [w_j^{m+1} - y_j^{m+1}(L_1)] Y_j(L_1), \quad j = 1, 2.$$

From (9.1.1) and (9.A.1.3) we have

$$p = \frac{\mu z_1 N_{i1}}{\xi z_2 N_{i2}}. \tag{9.A.1.5}$$

From (9.1.2) and (9.1.5) we have (9.2.5). From (9.2.2) and (9.2.5) we have (9.2.6).

Substituting (9.A.1.4) into $Y_1^* + Y_2^* = (m+1)F_{i1}/\xi\rho$ together with (9.A.1.5) yields

$$z_1^2 N_{s1} + pz_2^2 N_{s2} = \frac{\gamma z z_1 N_{i1}}{\xi}. \qquad (9.A.1.6)$$

Substituting $Y_j(L_1) = mR(L_1)/y_j^m(L_1)$ and Y_j^* into the two equations in (9.A.1.4) and then substituting $R(L_1)$ obtained from one equation to the other one, we obtain

$$p^{1-\gamma/\eta}\frac{N_{s1}}{N_{s2}} = \left(\frac{A_1}{A_2}\right)^{1/\eta}\left(\frac{z_1}{z_2}\right)\frac{z_1^{m+1} - y_1^{m+1}(L_1)}{w_2^{m+1} - y_2^{m+1}(L_1)} \equiv f(p) \qquad (9.A.1.7)$$

in which we use (9.A.1.6) and

$$\left(\frac{y_1(L_1)}{y_2(L_1)}\right)^m = \left(\frac{A_2}{A_1}\right)^{1/\eta}\left(\frac{p_{s1}}{p_{s2}}\right)^{\gamma/\eta}$$

which is obtained from (9.2.2). Using $L_1 = \Lambda(p)$, we can express (9.A.1.7) as a function of p. Substituting (9.A.1.7) into (9.A.1.6) yields

$$\frac{N_{i1}}{N_{s1}} = \xi\frac{z_1^2 f + p^{2-\gamma/\eta}z_2^2}{\gamma z z_1 f}. \qquad (9.A.1.8)$$

From $N_{i1} + N_{s1} = N_1$ and (9.A.1.8) we have

$$N_{i1} = \frac{\xi N_1(z_1^2 f + p^{2-\gamma/\eta}z_2^2)}{\gamma z z_1 f + \xi(z_1^2 f + p^{2-\gamma/\eta}z_2^2)},$$

$$N_{s1} = \frac{\gamma z z_1 f N_1}{\gamma z z_1 f + \xi(z_1^2 f + p^{2-\gamma/\eta}z_2^2)}. \qquad (9.A.1.9)$$

From $N_{i2} + N_{s2} = N_2$, (9.A.1.5) and (9.A.1.9) we have

$$N_{i2} = \frac{\mu z_1 N_{i1}}{\xi z_2 p}, \quad N_{s2} = N_2 - \frac{\mu z_1 N_{i1}}{\xi z_2 p}. \tag{9.A.1.10}$$

It is easy to check that from $w_1^m - y_1^m(L_1) = \upsilon m N_1 / Y_1$ in (9.A.1.2) and

$$\left[w_1^{m+1} - y_1^{m+1}(L_1)\right] Y_1(L_1) = \frac{(m+1) p_{s1} F_{s1}}{\gamma p}$$

in (9.A.1.4) we have

$$\frac{(m+1) z_1^2 N_{s1}}{\left[w_1^{m+1} - y_1^{m+1}(L_1)\right] \gamma p z} = \frac{\upsilon m N_1}{w_1^m - y_1^m(L_1)}.$$

Substituting the equation for N_{s1} in (9.A.1.9) into the above equation yields

$$M(p) \equiv \frac{(m+1) z_1^3 f}{\gamma z z_1 f + \xi \left(z_1^2 f + p^{2-\gamma/\eta} z_2^2\right)}$$
$$- \frac{\left[z_1^{m+1} - y_1^{m+1}(p)\right] \gamma p z}{w_1^m - y_1^m(p)} = 0. \tag{9.A.1.11}$$

As $0 < L_1 < L$, we see that p has to satisfy $0 < \Lambda(p) < L$ where $\Lambda(p)$ is defined in (9.2.5).

Substituting $Y_j(L_1) = m R(L_1) / y_j^m(L_1)$ into (9.A.1.2) yields

$$\left[\frac{z_1}{y_1(p)}\right]^m - 1 = \frac{\upsilon N_1}{R(L_1)},$$
$$\left[\frac{p z_2}{y_2(p)}\right]^m - 1 = \frac{\upsilon N_2}{R(L_1)}. \tag{9.A.1.12}$$

From the two equations in (9.A.1.2) and $N_1 + N_2 = N$, we have

$$N_1 = \left\{ \left(\frac{z_1}{y_1(p)}\right)^m - 1 \right\} N N_0(p),$$

$$N_2 = \left\{\left(\frac{pz_2}{y_2(p)}\right)^m - 1\right\} NN_0(p) \qquad (9.A.1.13)$$

where

$$N_0(p) \equiv \frac{1}{[pz_2/y_2(p)]^m + [z_1/y_1(p)]^m - 2}.$$

From the first equation in (9.A.1.12) we directly have

$$R(L_1) = \upsilon NN_0(p). \qquad (9.A.1.14)$$

By (9.2.1) we have

$$R(\omega_j) = \left[\frac{y_j(\omega_j)}{y_j(p)}\right]^m R(L_1), \quad j = 1, 2 \qquad (9.A.1.15)$$

in which $y_j(\omega_j)$ and $R(L_1)$ are functions of p.

We show how the 23 endogenous variables, N_{ij}, N_{sj}, N_j, F_{ij}, F_{sj}, w_j, p_{sj}, $(j = 1, 2)$, $c_1, c_2, c_s, n, R, p, L_1, U$, can be expressed as functions of the parameters. In Proposition 9.2.1 we solve p, p_{sj} and L_1. The wage rates w_j are given by (9.1.2), the urban population by (9.A.1.13), the labor division, N_{sj} and N_{ij}, by (9.A.1.9) and (9.A.1.10), the output of the industrial sectors F_{ij} by (9.1.1), the output of the service sectors F_{sj} by (9.1.4), the land rent at the boundary $R(L_1)$ by (9.A.1.14), the land rent distribution $R_j(\omega_j)$ by (9.A.1.15), the consumption components at any location, c_1, c_2, c_s, and L_h by (9.1.9), the utility level U by (9.1.7), the residential density at any location n by (9.1.10). We have thus proved Proposition 9.2.1.

10 Growth with International Trade and Urban Pattern Formation

So far we have been concerned with isolated economies. International trade was neglected in our analysis. The purpose of this chapter is to propose a simple two-country and one-commodity trade growth model with free capital mobility and urban pattern formation to provide some insights into relationships between commodity prices, factor prices, land values, production, preferences and trade volumes. The growth aspects of our model is based on the international macroeconomic one-sector growth model with perfect capital mobility. Irrespective of the analytical complexity of two-country, dynamic models with capital accumulation, many efforts have been made to examine the impact of savings, technology and various policies on trade patterns within this framework (e.g., Oniki and Uzawa, 1965, Bradhan, 1965, Rodriguez, 1975, Frenkel and Razin, 1987, Ruffin, 1979, Buiter, 1981, Fidlay, 1984, Rauch, 1991). Our modeling framework is based on the dynamic one-sector and two-country model (e.g., Wang, 1990, Ikeda and Ono, 1992, Devereux and Shi, 1991, Zhang, 1994c, 1995a, Turnovsky, 1997). We model behavior of households differently from these one-sector trade growth models in that we explicitly introduce consumption, savings and residential location in modeling decision makings of the households in the two countries.

We introduce land into our trade model. There are some factor-endowment models of international trade that have incorporated land as production input. For instance, Jones (1971) and Samuelson (1971) proposed a model in which one of two factors (land or capital) is used specially in the production of each of two commodities, while labor is used commonly in both production activities. This formulation has its origins in Ricardo's theory of rent and capital accumulation. Eaton (1987) extended this model to a dynamic framework with endogenous capital accumulation. Although our model deals with a similar issue, we consider land as a factor for housing consumption. We follow neo-classical urban economics to treat land mainly for amenity and housing uses.

The remainder of this chapter is organized as follows. Section 10.1 defines the basic model. Section 10.2 guarantees the existence of a unique equilibrium and the stability of the dynamic system. Sections 10.3 and 10.4 examine, respectively, effects of changes in country 1's propensity to hold wealth and country 2's propensity to hold wealth and working efficiency on the equilibrium structure of the world economy. Section 10.5 concludes the chapter. The appendix proves the conclusions in Section 10.2. This chapter is based on Zhang (1996b, 1998a, 1998b).

10.1 The Model

Similar to the trade models proposed by, for instance, Oniki and Uzawa (1965), Frenkel and Razin, (1987), and Ruffin (1979), we assume that the system consists of two countries, indexed by j, $j = 1, 2$ and only one good is produced in the system. The good is assumed to be composed of homogeneous qualities, and to be produced by employing two factors of production - labor and capital. Perfect competition is assumed to prevail in good markets both within each country and between the countries, and commodities are traded without any barriers such as transport costs or tariffs. We assume that there is no migration between the countries and labor markets are perfectly competitive within the country.

The population in each country is homogenous. The households achieve the same utility level regardless of where they locate. Each country is geographically linear and consists of two parts - the CBD and the residential area. Each country consists of a finite strip of land extending from a fixed central business district (CBD) with constant unit width. We assume that all economic activities are concentrated in the CBD. The residential area is occupied by households. We assume that the CBD is located at the left-side end of the linear territory.

The industrial production is similar to that in the one-sector neoclassical growth model. We assume that the industrial product can be either invested or consumed. We assume that the total labor force is fully employed by the production sector. We select the single good to serve as numeraire.

To describe the industrial sector, we introduce

N_j — the fixed population of country j, $j = 1, 2$;

$K_{ij}(t)$ — the capital stocks employed by country j's production sector at time t;

$w_j(t)$ and $r(t)$ — country j's wage rate and the rate of interest;

$F_j(t)$ and $C_j(t)$ — the output of country j's production sector and country j's net consumption level, respectively;

$K_j(t)$ — the capital stocks owned by country j;

$E(t) > 0 \ (< 0)$ — country 2's (1's) capital stocks utilized by country 1 (2); and

$K(t)$ — the world capital stock.

We assume that production is carried out by combination of capital and labor force in the form of

$$F_1 = zK_{i1}^\alpha N_1^\beta, \quad F_2 = K_{i2}^\alpha N_2^\beta, \quad \alpha + \beta = 1, \quad \alpha, \beta > 0 \qquad (10.1.1)$$

where α and β are parameters. Here, we call z the efficiency parameter of economic production. The parameter may be simply interpreted as a measurement of working efficiency difference between the two countries. If $z > 1$, we say that country 1's workers work more effectively than country 2's workers. The specified functional forms imply that the two countries have an identical production technology but different working efficiency. It should be remarked that it is conceptually possible to interpret z as a human capital index and thus extend our analysis to include dynamics of human capital accumulation.

The marginal conditions are given by

$$w_j = \frac{\beta F_j}{N_j}, \quad r = \frac{\alpha F_j}{K_{ij}}, \quad j = 1, 2. \qquad (10.1.2)$$

The world capital is equal to the sum of the capital stocks owned by the two countries, i.e.

$$K_1 + K_2 = K. \qquad (10.1.3)$$

According to the definitions of K_{ij}, K_j and E, we have

$$K_{i1} = K_1 + E, \quad K_{i2} = K_2 - E. \qquad (10.1.4)$$

The above equations state that the capital stocks utilized by each country is equal to the capital stocks owned by the country plus the foreign capital stocks.

We now describe housing production and behavior of households. First, we introduce

L_j — the fixed (territory) length of country j, $j = 1, 2$;

ω_j — the distance from the CBD to a point in the residential area in country j;

$R_j(\omega_j, t)$ — the land rent per household at location ω_j;

$k_j(\omega_j, t)$ and $S_j(\omega_j, t)$ — the capital stocks owned by and the savings made by the household at location ω_j, respectively;

$c_j(\omega_j,t)$ and $y_j(\omega_j,t)$ — the consumption and the net income of the household at location ω_j, respectively; and

$n_j(\omega_j,t)$ and $L_{hj}(\omega_j,t)$ — the residential density and the lot size of the household at location ω_j.

According to the definitions of L_{hj} and n_j, we have

$$n_j(\omega_j,t) = \frac{1}{L_{hj}(\omega_j,t)}, \quad 0 \leq \omega_j \leq L_j, \quad j = 1, 2. \tag{10.1.5}$$

We assume the public land ownership, which means that the revenue from land is equally shared between the population within each country. The total land revenue is given by

$$\overline{R}_j(t) = \int_0^L R_j(\omega,t) d\omega_j, \quad j = 1, 2. \tag{10.1.6}$$

The income from land per household is given by $\bar{r}_j(t) = \overline{R}_j(t)/N_j$, $j = 1, 2$. The net income $y_j(\omega_j,t)$ per household at location ω_j consists of three parts: the wage income, the income from land ownership and the interest payment for the household's capital stocks. The net income is thus given by

$$y_j(\omega_j,t) = r(t)k_j(\omega_j,t) + w_j(t) + \bar{r}_j(t), \quad j = 1, 2. \tag{10.1.7}$$

Many factors may affect residential location. For instance, interactions between the residential location and different space-related variables, such as transportation infrastructures, traffic congestion, traveling times to various facilities and working places, neighborhood amenity, racial prejudice, land rent and housing conditions, have been examined in urban economics. For simplicity of analysis, we take account of two spatial variables, local amenity $A_j(\omega_j,t)$ and leisure time $T_{hj}(\omega_j,t)$ in modeling the household's location decision.

In this chapter, we assume that households generally prefer a low-density residential area to a high one. We consider that the working time is fixed. Hence, the fixed available time, T_{0j}, is distributed between the leisure time and the time for traveling between the CBD and the dwelling site. For simplicity of analysis, we specify $A_j(\omega_j,t)$ and $T_{hj}(\omega_j,t)$ as follows

$$A_j(\omega_j,t) = \frac{\mu_{1j}}{n_j^{\mu_j}(\omega_j,t)}, \quad T_{hj}(\omega_j) = T_{0j} - \upsilon_j\omega_j,$$

$$\mu_{1j}, \mu_j, T_{0j}, \upsilon_j > 0. \qquad (10.1.8)$$

The function A_j implies that the amenity level at location ω_j is determined by the residential density at the location. The function T_{hj} means that the leisure time is equal to the total available time minus the traveling time. We assume that the traveling time between the CBD and the dwelling site is a linear function $\upsilon_{0j} + \upsilon_j\omega_j$ of the distance from the CBD to the dwelling site. We neglect possible impact of congestion and other factors on the traveling time.

We assume that utility level $U_j(\omega_j,t)$ of the household at location ω_j is dependent on the temporary consumption level $c_j(\omega_j)$, the lot size $L_{hj}(\omega_j)$, the leisure time $T_{hj}(\omega_j)$, the amenity level $A(\omega_j)$, and the household's wealth $k_j(\omega_j) + S_j(\omega_j) - \delta_k k_j(\omega_j)$, where δ_k ($1 > \delta_k > 0$) is the fixed depreciation rate of capital, as follows

$$U_j(\omega_j,t) = A_j T_{hj}^{\sigma_j} c_j^{\xi_j} L_{hj}^{\eta_j} \left(k_j + S_j - \delta_k k_j\right)^{\lambda_j},$$
$$\sigma_j, \xi_j, \eta_j, \lambda_j, \quad j = 1, 2 \qquad (10.1.9)$$

where σ_j, ξ_j, η_j, and λ_j are respectively interpreted as country j's propensities to use leisure time, to consume the commodity, to use lot size and to own wealth.

As the population is homogeneous within each country, we have

$$U_j(\omega_{j1},t) = U_j(\omega_{j2},t), \quad 0 \le \omega_{j1}, \omega_{j2} \le L_j. \qquad (10.1.10)$$

The budget constraint is given by

$$c_j(\omega_j,t) + R_j(\omega_j,t)L_{hj}(\omega_j,t) + S_j(\omega_j,t) = y_j(\omega_j,t). \qquad (10.1.11)$$

Maximizing U_j subject to the budget constraint yields

$$c_j(\omega_j) = \xi_j \rho_j y_j(\omega_j) + \xi_j \rho_{0j} y_j(\omega_j),$$
$$R_j(\omega_j) L_{hj}(\omega_j) = \eta_j \rho_j y_j(\omega_j) + \eta_j \rho_{0j} y_j(\omega_j),$$
$$S_j(\omega_j) = \lambda_j \rho_j y_j(\omega_j) - (\xi_j + \mu_j)\rho_{0j} k_j(\omega_j) \quad (10.1.12)$$

where

$$\rho_j = \frac{1}{\xi_j + \eta_j + \lambda_j}, \quad \rho_{0j} \equiv (1 - \delta_k)\rho_j, \quad j = 1, 2.$$

The above equations mean that the lot size and the consumption level are positively proportional to the net income and capital wealth, and the savings is positively proportional to the net income but negatively proportional to its capital stocks.

According to the definition of $S_j(\omega_j, t)$, we have the following capital accumulation for the household at location ω_j

$$\frac{dk_j(\omega_j)}{dt} = S_j(\omega_j) - \delta_{0j} k_j(\omega_j), \quad 0 \le \omega_j \le L_j.$$

Substituting $S_j(\omega_j)$ in (10.1.12) into the above equation yields

$$\frac{dk_j(\omega_j)}{dt} = s_j y_j(\omega_j) - \delta_j k_j(\omega_j), \quad 0 \le \omega_j \le L_j \quad (10.1.13)$$

where

$$s_j \equiv \lambda_j \rho_j, \quad \delta_j = \delta_0 + (\xi_j + \eta_j)\rho_{0j}.$$

As there is no migration, the following population constraints are held

$$\int_0^{L_j} n_j(\omega_j) d\omega_j = N_j, \quad j = 1, 2. \quad (10.1.14)$$

By the definitions of n_j, c_j, and C_j, we have

$$\int_0^{L_j} c_j(\omega_j) n_j(\omega_j) d\omega_j = C_j, \quad j = 1, 2. \tag{10.1.15}$$

The product is either invested or consumed, i.e.

$$F_1 + F_2 = C_1 + \overline{S}_1 + C_2 + \overline{S}_2 \tag{10.1.16}$$

where $\overline{S}_j(t)$ is country j's total savings, i.e.

$$\overline{S}_j \equiv \int_0^{L_j} S_j(\omega_j) n_j(\omega_j) d\omega_j, \quad j = 1, 2. \tag{10.1.17}$$

By the definitions of n_j, k_j and K_j, we have

$$\int_0^{L_j} k_j(\omega_j) n_j(\omega_j) d\omega_j = K_j, \quad j = 1, 2. \tag{10.1.18}$$

The total income $Y_j(t)$ of country j is equal to the sum of incomes of its population, i.e.

$$Y_j \equiv \int_0^{L_j} y_j(\omega_j) n_j(\omega_j) d\omega_j. \tag{10.1.19}$$

We have thus built the two-country trade model with economic growth and economic geography under perfectly competitive institution. The system has 18 space-time-dependent variables, k_j, c_j, L_{hj}, S_j, n_j, A_j, U_j, R_j, y_j ($j = 1, 2$), and 21 time-dependent variables, F_j, K_{ij}, K_j, Y_j, C_j, w_j, $\overline{S}_j$, $\overline{R}_j$, $\overline{r}_j$ ($j = 1, 2$), r, K, and E_j. It contains 39 independent equations. We now examine the properties of the dynamic system.

10.2 The Dynamic Properties of the Trade System

First, we show that the dynamics can be described by the motion of two variables, K_1 and K_2. By (10.1.7), (10.1.12) and (10.1.13) we see that the capital stocks

10.2 The Dynamic Properties of the Trade System

owned by per household and the net income are identical over space within each country at any point of time. Hence, we have

$$K_j = k_j N_j, \quad Y_j = y_j N_j, \quad Y_j = rK_j + w_j N_j + \overline{R}_j. \tag{10.2.1}$$

We rewrite the dynamics, (10.1.13), in aggregated terms as follows

$$\frac{dK_j}{dt} = s_j Y_j - \delta_j K_j, \quad j = 1, 2. \tag{10.2.2}$$

Our problem is to show that we can express $Y_j(t)$ as functions of K_1 and K_2.

Multiplying all the equations in (10.1.12) by $n_j(\omega_j, t)$ and then integrating the resulted equations from 0 to L_j with respect to ω_j, we obtain

$$C_j = \xi_j \rho_j Y_j + \xi_j \rho_{0j} K_j, \quad \overline{R}_j = \eta_j \rho_j Y_j + \eta_j \rho_{0j} K_j,$$
$$\overline{S}_j = \lambda_j \rho_j Y_j - (\xi_j + \mu_j) \rho_{0j} K_j, \quad j = 1, 2. \tag{10.2.3}$$

Substituting $\overline{R}_j$ in (10.2.1), and r and w_j in (10.1.2) into Y_j in (10.2.1), we have

$$Y_j = \frac{\alpha K_j F_j / K_{ij} + \beta F_j + \eta_j \rho_{0j} K_j}{1 - \eta_j \rho_j}. \tag{10.2.4}$$

From $r = \alpha F_1 / F_{i1} = \alpha F_2 / F_{i2}$ and (10.1.1), we have $K_{i1} / F_{i2} = z^{1/\beta} N_1 / N_2$. From this equation and (10.1.4), we have

$$E = z_1 K_2 - z_2 K_1, \quad K_{i1} = z_1 K, \quad K_{i2} = z_2 K \tag{10.2.5}$$

where

$$z_1 \equiv \frac{z^{1/\beta} N_1}{z^{1/\beta} N_1 + N_2}, \quad z_2 \equiv \frac{N_2}{z^{1/\beta} N_1 + N_2}.$$

We can thus express K_{ij} and F_j as functions of K_1 and K_2.

From the first equation in (10.2.5), we see that the foreign capital E is also a function of K_1 and K_2. The direction of trade is determined by the two countries' level, K_j/N_j of capital stocks owned by per capita and the working efficiency difference z. If the working force has the same level of working efficiency in the two countries, i.e., $z = 1$, and the capital stocks owned by per capita are equal between the two countries, then there is no trade (or trade is in balance), i.e., $E = 0$.

By (10.2.2), (10.2.5) and (10.2.4), we have

$$\frac{dK_1}{dt} = \frac{(\alpha K_1/z_1 K + \beta)\lambda_1 z z_1^\alpha N_1^\beta K^\alpha - (\xi_1 + \delta_k \lambda_1)K_1}{\xi_1 + \lambda_1},$$

$$\frac{dK_2}{dt} = \frac{(\alpha K_2/z_2 K + \beta)\lambda_2 z_2^\alpha N_2^\beta K^\alpha - (\xi_2 + \delta_k \lambda_2)K_2}{\xi_2 + \lambda_2}. \quad (10.2.6)$$

As $K = K_1 + K_2$, we see that the dynamic system, (10.2.6), are only dependent on $K_1(t)$ and $K_2(t)$. Accordingly, the about two differential equations determine the capital stocks owned by the two countries, independent of the other variables in the system. We can show that all the other variables in the system are uniquely determined as functions of K_j and ω_j ($0 \le \omega_j \le L_j, j = 1, 2$) at any point of time. The following proposition is proved in the appendix.

Proposition 10.2.1.
For any given (positive) level of the capital stocks, $K_1(t)$ and $K_2(t)$, at any point of time, all the other variables in the system are uniquely determined as functions of $K_1(t)$ and $K_2(t)$. The dynamics of $K_1(t)$ and $K_2(t)$ are determined by (10.2.6).

We can thus explicitly determine the motion of the system over time and space. It should be remarked that the conclusion that the dynamics can be explicitly given as in (10.2.6) is important. It makes it possible to determine stability of the system.

Before further analyzing the dynamic properties of the system, we examine how the differences in values of some variables between the two countries are determined at any point of time. Substituting K_{i1} and K_{i2} in (10.2.5) into (10.1.1) and (10.1.2), we have

10.2 The Dynamic Properties of the Trade System

$$\frac{F_1}{F_2} = \frac{w_1}{w_2} = z^{1/\beta}, \quad r = \alpha z \left(\frac{N_1}{z_1 K}\right)^{\beta} \tag{10.2.7}$$

where we assume $N_1 = N_2$. If country 1 works more effectively than country 2, both country 1's output and wage rate are higher than country 2. In the free trade system, if the world capital stocks K is increased, the interest rate r is reduced. From (10.A.1.3) and (10.A.1.2), we get the ratio of the residential densities as follows

$$\frac{n_1(0)}{n_2(0)} = \frac{\upsilon\left[1 - (1 - \upsilon_2 L_2 / T_{02})^{B_2}\right]}{\left[1 - (1 - \upsilon_1 L_1 / T_{01})^{B_1}\right]},$$

$$\frac{n_1(\omega_1)}{n_2(\omega_2)} = \frac{n_1(0)(1 - \upsilon_1 \omega_1 / T_{01})^{B_1 - 1}}{n_2(0)(1 - \upsilon_2 \omega_2 / T_{02})^{B_2 - 1}}, \quad 0 \leq \omega_j \leq L_j, \quad j = 1, 2$$

$$\tag{10.2.8}$$

where

$$\upsilon \equiv \frac{\upsilon_1 B_1 N_1 T_{01}}{\upsilon_2 B_2 N_2 T_{02}}, \quad B_j \equiv \frac{\sigma_j}{\mu_j + \eta_j} + 1.$$

We see that the ratios are determined by the differences in the population, the transportation conditions, the available times, the crowding effects, the preferences and the territory sizes between the two countries. For instance, when the two countries are identical in the transportation condition, the available time, the population, the residential distribution is different if the two countries have different preferences, i.e., $B_1 \neq B_2$. It should be remarked that as we have already explicitly solved all the variables as functions of $K_1(t)$ and $K_2(t)$ at any point of time, it is direct to compare all the variables in the system between the two countries.

We now examine whether or not the dynamic system has equilibrium. By (10.2.6) equilibrium of the dynamic systems is given by

$$\left(\frac{\alpha K_1}{z_1 K} + \beta\right) \lambda_1 z z_1^{\alpha} N_1^{\beta} K^{\alpha} = (\xi_1 + \delta_k \lambda_1) K_1,$$

$$\left(\frac{\alpha K_2}{z_2 K} + \beta\right) \lambda_2 z_2^{\alpha} N_2^{\beta} K^{\alpha} = (\xi_2 + \delta_k \lambda_2) K_2. \tag{10.2.9}$$

176   10   Growth with International Trade and Urban Pattern Formation

It can be shown that the above equations have a unique solution.

Proposition 10.2.2.
The dynamic trade system has a unique stable equilibrium.

The above proposition is proved in Appendix 10.A.2. The ratio of the capital stocks owned by the two countries and the foreign capital stocks E at equilibrium are given by

$$\frac{K_1}{K_2} = \Lambda \equiv \Lambda_1 + \left(\Lambda_1^2 + \Lambda_0\right)^{1/2}, \quad E = \left(1 - \frac{N_2 \Lambda}{z^{1/\beta} N_1}\right) z_1 K_2 \quad (10.2.10)$$

where

$$\Lambda_0 \equiv \frac{(\xi_2/\lambda_2 + \delta_k) z^{1/\beta} N_1}{(\xi_1/\lambda_1 + \delta_k) N_2}, \quad \Lambda_1 \equiv \frac{\alpha/\beta z_1 + 1}{2} \Lambda_0 - \frac{\alpha/\beta z_2 + 1}{2}$$

(10.2.11)

where we use (10.A.2.1) and the definitions of z_1 and z_2. Here, if $\xi_1/\lambda_1 > (<) \xi_2/\lambda_2$, we say that country 1's net propensity to own wealth is lower (higher) than country 2. From (10.1.9), we see that when $\xi_1/\lambda_1 > \xi_2/\lambda_2$, country 1's propensity to own wealth is lower than country 2. From (10.2.10), we see that it is not easy to explicitly determine the sign of E. The trade direction is affected by the differences in the population, preferences and working efficiency between the two countries. To examine the sign of E, we examine a few special cases. In the case that the two countries have an identical working efficiency and net propensity to own wealth, i.e., $z = 1$ and $\xi_1/\lambda_1 = \xi_2/\lambda_2$, we have: $K_1/K_2 = N_1/N_2$ and $E = 0$. The country with larger population has more capital stocks but trade is in balance.

We now examine the case that the two countries have identical population and working efficiency and country 1's net propensity to own wealth is higher than country 2, i.e., $N_1 = N_2$, $z = 1$, and $\xi_1/\lambda_1 < \xi_2/\lambda_2$. In this case, we have

$$\Lambda_0 = \frac{\xi_2/\lambda_2 + \delta_k}{\xi_1/\lambda_1 + \delta_k} > 1, \quad \Lambda_1 = \frac{(1+\alpha)(\Lambda_0 - 1)}{2\beta} > 0.$$

By (10.2.10), we have

10.3 Country 1's Propensity to Own Wealth

$$\frac{K_1}{K_2} = \Lambda_1 + \left(\Lambda_1^2 + \Lambda_0\right)^{1/2} > 1, \quad E = (1-\Lambda)z_1 K_2 < 0. \tag{10.2.12}$$

When country 1's net propensity to own wealth is higher than country 1, country 1 has more capital stocks than country 2 and some of country 1's capital stocks are utilized by country 2.

Another case is that the two countries have identical net propensity to own wealth and identical population, but different working efficiency, i.e., $N_1 = N_2$, $\xi_1/\lambda_1 = \xi_2/\lambda_2$ and $z \neq 1$. As $\Lambda_0 = z^{1/\beta}$ and $\Lambda_1 = (z^{1/\beta} - 1)/2$, the country which has a higher working efficiency owns more capital stocks and some of its capital stocks is utilized by the other country, i.e., $K_1/K_2 > (<) 1$ and $E < (>) 0$ in the case of $z > (<) 1$. In the case that the two countries have identical population and country 1 works more effectively than country 2 and country 1's net propensity to own wealth is higher than country 2, i.e., $N_1 = N_2$, $\xi_1/\lambda_1 < \xi_2/\lambda_2$ and $z > 1$, it is direct to check: $\Lambda_1 > 0$, $\Lambda_0 > 1$, $K_1/K_2 > 1$ and $E < 0$.

We now examine effects of changes in some parameters on the equilibrium structure of the world economy. In the remainder of this chapter, we assume: $N \equiv N_1 = N_2$.

10.3 Country 1's Propensity to Own Wealth

This section examines effects of changes in country 1's propensity λ_1 to own wealth on the trade system. Taking derivatives of (10.A.2.2) and (10.2.10) with respect to λ_1, we obtain

$$\frac{\beta}{K}\frac{dK}{d\lambda_1} = \frac{\xi_1}{(\xi_1 + \delta_k \lambda_1)\lambda_1} - \frac{\beta}{(\alpha\Lambda/z_1 + \beta\Lambda + \beta)\Lambda}\frac{d\Lambda}{d\lambda_1},$$

$$\frac{d(K_1/K_2)}{d\lambda_1} = \frac{d\Lambda}{d\lambda_1} = \frac{d\Lambda_1}{d\lambda_1} + \frac{2\Lambda_1 d\Lambda_1/d\lambda_1 + d\Lambda_0/d\lambda_1}{2(\Lambda_1^2 + \Lambda_0)^{1/2}} > 0$$

(10.3.1)

where

$$\frac{d\Lambda_0}{d\lambda_1} = \frac{\xi_1\Lambda_0}{(\xi_1 + \delta_k\lambda_1)\lambda_1} > 0, \quad \frac{d\Lambda_1}{d\lambda_1} = \frac{\alpha/\beta z_1 + 1}{2}\frac{d\Lambda_0}{d\lambda_1} > 0. \quad (10.3.2)$$

The world capital stocks may be either increased or reduced. The ratio K_1/K_2 of the two countries' capital stocks is increased. This implies that the difference in the two countries' capital stocks is enlarged as country 1's propensity to own wealth is increased. From (10.3.1) and (10.3.2), we see that it is difficult to explicitly interpret the results as the expressions are too complicated. For convenience of interpretation, we further require that the two countries have the same level of working efficiency and country 1's net propensity to own wealth is higher than country 2, i.e., $z=1$ and $\xi_1/\lambda_1 < \xi_2/\lambda_2$. From the discussion in the preceding section, we have

$$\Lambda_0 = \frac{\xi_2/\lambda_2 + \delta_k}{\xi_1/\lambda_1 + \delta_k} > 1, \quad \Lambda_1 = \frac{(1+\alpha)(\Lambda_0 - 1)}{2\beta} > 0.$$

We determine K_1/K_2 and E by (10.2.12).

Taking derivatives (10.A.2.3) with respect to λ_1 yields

$$\frac{\beta}{K_1}\frac{dK_1}{d\lambda_1} = \frac{\xi_1}{(\xi_1 + \delta_k\lambda_1)\beta\lambda_1} + \frac{\alpha\Lambda^*(\Lambda - \alpha)}{(1+\Lambda)\Lambda}\frac{d\Lambda}{d\lambda_1} > 0,$$

$$\frac{\beta}{K_2}\frac{dK_2}{d\lambda_1} = \frac{\xi_1}{(\xi_1 + \delta_k\lambda_1)\beta\lambda_1} - \left(\frac{\beta\Lambda^*}{\Lambda} + \frac{1}{1+\Lambda}\right)\frac{d\Lambda}{d\lambda_1} \quad (10.3.3)$$

where we use $dK/d\lambda_1$ given by (10.3.1) and

$$\Lambda^* \equiv \frac{1}{\alpha\Lambda + \Lambda + \beta}.$$

Country 1's capital stocks is increased as its propensity to own wealth is increased. The impact on country 2's capital stocks is uncertain. If the world capital stocks is reduced, then country 2's capital stocks is certainly decreased. Even when K is increased, K_2 may still be decreased. From (10.2.5) and (10.3.3), we have

$$2\frac{dE}{d\lambda_1} = \frac{(K_2 - K_1)\xi_1}{(\xi_1 + \delta_k\lambda_1)\beta\lambda_1}$$

10.3 Country 1's Propensity to Own Wealth

$$-\left(\frac{\beta\Lambda^*K_2}{\Lambda} + \frac{K_2}{1+\Lambda} + \frac{\alpha\Lambda^*(\Lambda-\alpha)}{(1+\Lambda)\Lambda}\right)\frac{d\Lambda}{d\lambda_1},$$

$$2\frac{dK_{i1}}{d\lambda_1} = 2\frac{dK_{i2}}{d\lambda_1} = \frac{dK}{d\lambda_1} \quad (10.3.4)$$

where we use $z_1 = z_2 = 1/2$ and $K_2 < K_1$. The trade gap between the two countries is enlarged as country 1 increases its propensity to own wealth. The capitals utilized by the two countries are increased (reduced) if the world capital stocks are increased (reduced). From (10.1.1) and (10.1.2), we directly have

$$\frac{1}{F_1}\frac{dF_1}{d\lambda_1} = \frac{1}{F_2}\frac{dF_2}{d\lambda_1} = \frac{1}{w_1}\frac{dw_1}{d\lambda_1} = \frac{1}{w_w}\frac{dw_2}{d\lambda_1} = \frac{\alpha}{K}\frac{dK}{d\lambda_1},$$

$$\frac{1}{r}\frac{dr}{d\lambda_1} = -\frac{\beta}{K}\frac{dK}{d\lambda_1}. \quad (10.3.5)$$

Country 1's propensity to own wealth is increased (reduced), the levels of output and wage rates of the two countries are increased (reduced) and the rate of interest is decreased (increased). From

$$Y_j = \frac{\delta_j K_j}{s_j} = \left(\delta_k + \frac{\xi_j + \eta_j}{\lambda_j}\right)K_j$$

we have

$$\frac{1}{K_1}\frac{dY_1}{d\lambda_1} = \frac{(\alpha\xi_1 - \beta\delta_k\lambda_1)(\xi_1 + \eta_1)}{(\xi_1 + \delta_k\lambda_1)\lambda_1^2\beta} + \frac{\delta_k\xi_1}{(\xi_1 + \delta_k\lambda_1)\beta\lambda_1}$$

$$+ \frac{\alpha\delta_1\Lambda^*(\Lambda-\alpha)}{(1+\Lambda)s_1\Lambda}\frac{d\Lambda}{d\lambda_1},$$

$$\frac{dY_2}{d\lambda_1} = \frac{\delta_2}{s_2}\frac{dK_2}{d\lambda_1}. \quad (10.3.6)$$

The income of country 1 is increased (we require $\alpha\xi_1 \geq \beta\delta_k\lambda_1$), but that of country 2 may be either increased or decreased. From $Y_j = \delta_j K_j / s_j$ and (10.2.3), we get the effects on the land revenue and consumption levels as follows

$$\frac{\lambda_1}{\eta_1}\frac{d\overline{R}_1}{d\lambda_1} = \frac{\lambda_1}{\xi_1}\frac{dC_1}{d\lambda_1} = \frac{(\alpha\xi_1 - \beta\delta_k\lambda_1)K_1}{(\xi_1 + \delta_k\lambda_1)\lambda_1\beta} + \frac{\alpha K_1\Lambda^*(\Lambda - \alpha)}{(1+\Lambda)\Lambda}\frac{d\Lambda}{d\lambda_1},$$

$$\frac{\lambda_2}{\eta_2}\frac{d\overline{R}_2}{d\lambda_1} = \frac{\lambda_2}{\xi_2}\frac{dC_2}{d\lambda_1} = \frac{dK_2}{d\lambda_1}. \qquad (10.3.7)$$

Country 1's land income and consumption level are increased. The effects on country 2's land income and consumption level are the same as on the world capital stocks. From (10.1.12), we get that country 1's land rent at any location is increased, i.e., $dR_1(\omega_1)/d\lambda_1 > 0$; and the sign of $dR_2(\omega_2)/d\lambda_1$ is the same as that of $dK/d\lambda_1$.

Summarizing the discussion in this section, we have the following proposition.

Proposition 10.3.1.
Assume that the two countries have identical population and working efficiency and country 1's net propensity to own wealth is higher than country 2, i.e., $N_1 = N_2$, $z = 1$ and $\xi_1/\lambda_1 < \xi_2/\lambda_2$. Then country 1 owns more capital stocks than country 2 and some of country 1's capital stocks are employed by country 2, i.e., $K_1/K_2 > 1$ and $E < 0$. An increase in country 1's propensity to own wealth λ_1, has the following impact on the system: 1) the difference in capital stocks and trade gap are enlarged, $d(K_1/K_2)/d\lambda_1 > 0$ and $dE/d\lambda_1 < 0$; 2) country 1's capital stocks K_1, total income Y_1, the level of consumption C_1, the land revenue $\overline{R}_1$, land rent $R_1(\omega_1)$ at any location, are increased; 3) the world capital stocks K may be either increased or reduced; if $dK/d\lambda_1 < (>)\, 0$, then 4) the outputs F_1 and F_2, the wage rates w_1 and w_2, and capital stocks K_{i1} and K_{i2}, employed by the two countries, are increased (reduced), but the rate of interest is reduced (increased); 5) country 2's capital stocks K_2, total income Y_2, the level of consumption C_2, the land revenue $\overline{R}_2$, land rent $R_2(\omega_2)$ at any location, are increased (reduced).

We will not interpret how a shift in the parameter changes the variable values in details as this requires examining how all the equations in the system are affected.

10.4 Country 1's Working Efficiency

This section discusses the effects of changes in country 1's working efficiency z on the equilibrium structure. We only examine the impact on the capital stocks. It is direct to obtain the effects on the other variables similarly as in the preceding section. Taking derivatives of (10.A.2.2) and (10.2.10) with respect to z yield

$$\frac{\beta}{K}\frac{dK}{dz} = \frac{1+\beta z^{1/\beta}}{(1+z^{1/\beta})\beta z} - \frac{\alpha \Lambda^2 / z^{1+1/\beta} + \beta d\Lambda/dz}{(\alpha \Lambda/z_1 + \beta \Lambda + \beta)\Lambda},$$

$$\frac{d(K_1/K_2)}{dz} = \frac{d\Lambda}{dz} = \frac{d\Lambda_1}{dz} + \frac{2\Lambda_1 d\Lambda_1/dz + d\Lambda_0/dz}{(\Lambda_1^2 + \Lambda_0)^{1/2}} > 0 \quad (10.4.1)$$

where

$$\frac{d\Lambda_0}{dz} = \frac{\Lambda_0}{\beta z}, \quad \frac{d\Lambda_1}{dz} = \frac{\Lambda_0/z - z^{\alpha/\beta}}{2\beta^2}\alpha + \frac{\Lambda_0}{2\beta z}. \quad (10.4.2)$$

in which $\Lambda_0/z > z^{\alpha/\beta}$. By (10.A.2.3) we have

$$\frac{1}{K_1}\frac{dK_1}{d\lambda_1} = \frac{1}{K}\frac{dK}{dz} + \frac{1}{(1+\Lambda)\Lambda}\frac{d\Lambda}{dz},$$

$$\frac{1}{K_2}\frac{dK_2}{dz} = \frac{1}{K}\frac{dK}{dz} - \frac{1}{1+\Lambda}\frac{d\Lambda}{dz}. \quad (10.4.3)$$

We see that the effects of changes in z on the capital stocks are quite similar to the effects of changes in λ_1.

10.5 On International Trade and Spatial Structures

The chapter proposed a two-country trade model with endogenous capital accumulation and economic geography. We showed how differences in preferences, population and territory sizes affect the trade patterns and the economic geography in the dynamic competitive world economy. We proved that the dynamic trade system has a unique stable equilibrium. We also examined the effects of changes in the two countries' propensities to save and working efficiency difference on the equilibrium structure of the world economy.

As the model is built on the basis of different neoclassical schools, it is conceptually possible to extend it according to the literature on neoclassical growth theory, neoclassical dynamic trade theory, and neoclassical urban economics. It is quite important direction to extend our model is to introduce endogenous knowledge.

Appendix

A.10.1 Proving Proposition 10.2.1

We already uniquely determined k_j, y_j, E, K, F_j, K_{ij} and Y_j as functions of K_1 and K_2. From (10.2.3) and (10.1.12), we directly determine $\overline{R}_j$, C_j, $\overline{S}_j$, c_j and S_j. The income from land ownership per household is given by: $\bar{r} = \overline{R}/N$. The rate of the interest r, and the wage rate w_j are uniquely determined by (10.1.2).

Substituting (10.1.5) and (10.1.8) into $U_j(\omega_j)$ in (10.1.9) yields

$$U_j(\omega_j) = \mu_{1j}\left(T_{0j} - \upsilon_j\omega_j\right)^{\sigma_j} c_j^{\xi_j} n_{hj}^{-\mu_j-\eta_j}\left(k_j + S_j - \delta_k k_j\right)^{\lambda_j}.$$

(10.A.1.1)

Substituting (10.A.1.1) into $U_j(0) = U_j(\omega_j)$, we have

$$n_j(\omega_j) = n_j(0)\left(1 - \frac{\upsilon_j\omega_j}{T_{0j}}\right)^{\sigma_j/(\mu_j+\eta_j)}.$$

(10.A.1.2)

Substituting (10.A.1.2) into (10.1.14) and then integrating the resulted equation from 0 to L_j, we have

$$n_j(0) = \frac{\upsilon_j B_j N_j T_{0j}^{-1}}{1 - \left(1 - \upsilon_j L_j / T_{0j}\right)^{B_j}}, \quad j = 1, 2.$$

(10.A.1.3)

We assume $1 - \upsilon_j L_j / T_{0j} > 0$. This simply means that the available time is sufficient to travel from the boundary of the country to the CBD. By (10.A.1.2) and (10.A.1.3), we determine the residential density at any location in the two countries.

The lot size per household is given by: $L_{hj}(\omega_j) = 1/n_j(\omega_j)$. From (10.1.12), we have the land rent $R_j(\omega_j)$. The local amenity $A_j(\omega_j)$ is given by (10.1.8). The utility level $U_j(\omega_j)$ is given by (10.1.9).

We thus showed how to determine all the variables in the system as unique functions of K_1 and K_2.

A.10.2 Proving Proposition 10.2.2.

First, we show that the system has a unique equilibrium. Dividing the first equation by the second one in (10.2.9), we have

$$\frac{K_1}{K_2} = \Lambda \tag{10.A.2.1}$$

where Λ is defined in (10.2.10). Substituting (10.A.2.1) into the first equation in (10.2.9), we have

$$K = \overline{\Lambda} \equiv \left[\frac{\alpha/z_1 + \beta/\Lambda + \beta}{\xi_1 + \delta_k \lambda_1} \lambda_1 z z_1^\alpha N_1^\beta \right]^{1/\beta} \tag{10.A.2.2}$$

where $\overline{\Lambda}$ is a constant. By $K_1 + K_2 = K$, (10.A.2.1) and (10.A.2.2), we have

$$K_1 = \frac{\Lambda \overline{\Lambda}}{1 + \Lambda}, \quad K_2 = \frac{\overline{\Lambda}}{1 + \Lambda}. \tag{10.A.2.3}$$

We obtained a unique equilibrium. We now provide stability conditions for the equilibrium.

It is easy to calculate that the two eigenvalues, ϕ_1 and ϕ_2, are given by

$$\phi^2 - (a_1 + b_1)\phi + a_1 b_2 - a_2 b_1 = 0 \tag{10.A.2.4}$$

where

$$a_1 \equiv - \frac{\lambda_1 z z_1^\alpha K_2 N_1^\beta K^{-\beta}/K_1^2 + (\xi_1 + \delta_k \lambda_1)/K}{\xi_1 + \lambda_1} \beta K_1 < 0,$$

$$a_2 \equiv \frac{\lambda_1 z z_1^\alpha N_1^\beta K^{-\beta}/K_1 - (\xi_1 + \delta_k \lambda_1)/K}{\xi_1 + \lambda_1} \beta K_1,$$

$$b_1 \equiv \frac{\lambda_2 z z_2^\alpha N_2^\beta K^{-\beta}/K_2 - (\xi_2 + \delta_k \lambda_2)/K}{\xi_2 + \lambda_2} \beta K_2,$$

$$b_2 \equiv -\frac{\lambda_2 z z_2^\alpha K_1 N_2^\beta K^{-\beta}/K_2^2 + (\xi_2 + \delta_k \lambda_2)/K}{\xi_2 + \lambda_2} \beta K_2 < 0. \quad (10.A.2.5)$$

As a_1 and b_2 are negative, we see that the system is stable if $a_2 b_1 - a_1 b_2 < 0$. From (10.A.2.5), we directly obtain

$$a_2 b_1 - a_1 b_2 =$$
$$\frac{\lambda_1 z z_1^\alpha N_1^\beta K^{1+\alpha}(\xi_2 + \delta_k \lambda_2)/K_1^2 + \lambda_2 z z_2^\alpha N_2^\beta K^{1+\alpha}(\xi_1 + \delta_k \lambda_1)/K_2^2}{(\xi_1 + \lambda_1)(\xi_2 + \lambda_2)K^2}.$$

(10.A.2.6)

Accordingly, the unique equilibrium is stable.

We thus proved Proposition 10.2.2.

11 Nonlinear Dynamics of a Multi-City System

It may be said that we are in the midst of a profound rupture between older and emergent notions of scientific explanation (e.g., Haken, 1977, 1983, Nicolis and Prigogine, 1977, Prigogine, 1980, 1997, Prigogine and Stengers, 1984). The revolution has been much influenced by developments in nonlinear dynamics and computation in the past few decades. Stability and instability, order and chaos, determinism and stochasticity have been reconceptualized. This may be witnessed by about sixty volumes of the Series in Synergetics edited by Haken. Synergetics, a name for nonlinear interactive dynamics, deals with systems composed of many parts, interacting with each other over time and space. It may be considered as a strategy for coping with complex systems, by focusing on those situations where these systems qualitatively change their macroscopic behavior.

Social scientists have turned their prodigious efforts into assimilating some of the new doctrines and mathematical techniques. Inspired by the pathbreaking works in nonlinear dynamic theory, social scientists have attempted to explain complicated dynamic phenomena such as catastrophes, bifurcations, fractals and chaos within endogenous deterministic frameworks. One of the important topics of applying nonlinear dynamic theory to social sciences is to explain the rise and fall of cities and regions (e.g., Allen and Sanglier, 1979, Wilson, 1981, Andersson, 1986, Arthur, 1989, Dendrinos and Sonis, 1990, Puu, 1989, Weidlich and Haag, 1983, Nijkamp and Reggiani, 1992, and Rosser, 1991).

This chapter provides an example of urban economic dynamics to show how one can explain urban complexity, while still working within the framework of traditional urban economic theory but by applying nonlinear dynamic methods. There are numerous factors involved in such an endeavor (such as creation, diffusion and utilization of knowledge; infrastructures; and endogenous growth and net in-migration of population) which may result in scale and scope economies. Rather than taking account of all these important factors, we are only concerned here with the dynamics of endogenous population growth and capital accumulation. For simplicity, the focus is first on an isolated island economy. Then, essential forces of various factors in forming urban pattern are addressed. Finally, the isolated economy is extended to an economic system consisting of interacting multiple islands. This chapter is based on Zhang (1994a, 1994b).

11.1 An Isolated Island Economy

Consider an island with fixed radius. Let us assume that all the economic activities are located at the center of this island. The system is assumed to be isolated in the sense that there is neither migration nor trade with the outside world. By isolating the system, the focus is on the endogenous mechanism of urban growth and pattern formation. The system is geographically similar to the model in Chapter 2. The island consists of a city with a residential area surrounding the location of a single production site, i.e., the city is monocentric. It has a single pre-specified center of fixed size called the central business district (CBD). The economic system consists of two sectors - industry and service. The industry supplies goods either for consumption or for investment.

We select the industrial product to serve as the numeraire, i.e., with all the other prices being measured relative to its price. Wage rates are identical among different professions due to perfect competition in the labor market. We neglect time and costs needed for professional transformation. We assume that capital is freely mobile between the two sectors due to the perfectly competitive mechanism. This implies that the interest rate is identical between the two sectors. The assumption implies that the capital market is at its temporary equilibrium at any point of time. As we use a single kind of capital and assume the capital market to be well informed, this assumption is reasonable. In reality, there are multiple kinds of physical capital (plants and equipment) and demand for and supply of each kind of capital stock are rather complex.

We use subscript indexes i and s to denote the industrial sector and the service sector, respectively, and introduce the following variables

L — the distance from the CBD to the perimeter of the island;
ω — the distance from the CBD to a point in the residential area;
$R(\omega,t)$ — the land rent at location ω at time t;
$N(t)$ — the total labor force at time t;
$K(t)$ — the total capital stock;
$N_i(t)$ and $K_i(t)$ — the labor force and capital stock employed by the industrial sector, respectively;
$N_s(t)$ and $K_s(t)$ — labor force and capital stock employed by the service sector, respectively;
$F_i(t)$ and $F_s(t)$ — the output levels of the industrial and service sectors, respectively;
$C_i(t)$ and $C_s(t)$ — the total consumption levels of industrial good and services by the population at time t, respectively;

$p(t)$, $w(t)$ and $r(t)$ — the price of services, wage rate and rate of interest, respectively.

We now describe the production and consumption of the island economy.

Production, capital and labor markets
We assume that industrial and service production are achieved by combinations of capital and labor as follows

$$F_i = K_i^{\alpha_i} N_i^{\beta_i}, \quad F_s = K_s^{\alpha_s} N_s^{\beta_s}, \quad \alpha_i + \beta_i = 1, \quad \alpha_s + \beta_{si} = 1,$$
$$\alpha_i, \beta_i, \alpha_s, \beta_s > 0. \qquad (11.1.1)$$

Since services are consumed simultaneously as they are produced (i.e., there is no excess demand or supply in this sector), we have

$$C_s(t) = F_s(t). \qquad (11.1.2)$$

The marginal conditions for efficient production in the two sectors are given by

$$r = \frac{\alpha_i F_i}{K_i} = \frac{\alpha_s p F_s}{K_s}, \quad w = \frac{\beta_i F_i}{N_i} = \frac{\beta_s p F_s}{N_s}. \qquad (11.1.3)$$

Consumption choices
We assume that a homogeneous group of residents reside in the island; a household's utility level, reflecting its consumption of industrial goods, services and housing, can be represented in the following Cobb-Douglas log-linear form

$$U(\omega,t) = U[c_s(\omega,t), c_i(\omega,t), L_h(\omega,t)] = c_s^\gamma c_i^\xi L_h^\eta,$$
$$1 > \gamma, \xi, \eta > 0 \qquad (11.1.4)$$

where $c_s(\omega,t)$, $c_i(\omega,t)$, and $L_h(\omega,t)$ are the consumption levels of services, industrial good and housing of a household at location ω and at time t, respectively. It should be remarked that housing production and the housing market in this chapter are extremely simplified. Housing has a set of intrinsic properties which make it significantly different from any other goods as mentioned in Chapter 2. There is a large literature regarding the complexity of housing dynamics (e.g., Sweeney, 1974, Arnott, 1987, Fujita, 1982, Brueckner, 1981, Brueckner and Rabenau, 1981, Hockman and Pines, 1980a, 1980b among many others); this work enables one further extend our simple framework.

188   11   Nonlinear Dynamics of a Multi-City System

For simplicity, we specify parameter values of γ, ξ and η by assuming that $\gamma + \xi = \eta$. It can be shown that this assumption can be easily relaxed. Let Y denote the net income of the population and c the consumption rate of Y. The consumption budget of an individual is defined by cY/N. The consumer problem is then defined by

$$\max\ U(\omega,t),\ \ \text{s.t:}\ \ c_i(\omega) + pc_s(\omega) + R(\omega)L_h(\omega) = \frac{cY}{N} - \Gamma(\omega) \tag{11.1.5}$$

where $\Gamma(\omega)$ is the total cost of travel (in terms of the industrial good) between dwelling site w and the CBD. For simplicity, we specify $\Gamma(\omega)$ by

$$\Gamma(\omega) = \upsilon_0 \omega^\upsilon,\ \ \upsilon_0 > 0,\ \ 0 \leq \upsilon \leq 1. \tag{11.1.6}$$

Here, for convenience of analysis, we neglect any urban externalities, such as traffic congestion and crowding effects, which are important in shaping urban form (land use densities) and spatial land rent distribution. It is conceptually not difficult to take account of certain types of externalities in our model (e.g., Straszheim, 1987, Dendrinos, 1992, Mills and Hamilton, 1985), such as congestion, pollution, or densities.

The unique optimal solution for the typical resident in the city is given by

$$c_i = \xi s_0 y,\ \ c_s = \frac{\gamma s_0 y}{p},\ \ L_h = \frac{\eta s_0 y}{R} \tag{11.1.7}$$

in which

$$s_0 \equiv \frac{c}{\lambda + \xi + \eta},\ \ y(\omega) = \frac{cY}{N} - \Gamma(\omega),\ \ Y = F_i + pF_s.$$

Let $n(\omega,t)$ represent the residential density at dwelling site ω. According to the definitions, we have

$$n(\omega,t) = \frac{1}{L_h(\omega,t)},\ \ 0 \leq \omega \leq L,\ t \geq 0. \tag{11.1.8}$$

The population constraint is given by

11.1 An Isolated Island Economy

$$2\pi \int_0^L n(\omega)\omega\, d\omega = N. \quad (11.1.9)$$

The consumption constraints are given by

$$2\pi \int_0^L c_i(\omega)n(\omega)\omega\, d\omega = C_i, \quad 2\pi \int_0^L c_s(\omega)n(\omega)\omega\, d\omega = C_s. \quad (11.1.10)$$

Savings and capital formation
Capital accumulation according to standard theory is formulated by

$$\frac{dK}{dt} = sY - \delta K \quad (11.1.11)$$

where $s\ (\equiv 1 - c)$ is the savings rate and δ is the given depreciation rate of capital.

Population dynamics
On the basis of endogenous population theory (e.g., Haavelmo, 1954, Niehans, 1963, Pitchford, 1974, Becker, 1981), we suggest the following population dynamics

$$\frac{dN}{dt} = nN\left[C^q - \theta N\left(\frac{K}{N}\right)^m\right] \quad (11.1.12)$$

where $C\ (\equiv cY)$ is the consumption level, and n, q, θ and m are non-negative parameters. We interpret C^q as the capacity for supporting the population. This implies that the population which can suitably live (or survive) in a society is dependent upon the current consumption. The term $\theta N(K/N)^m$ expresses how wealth, acting as a checking force, affects population growth. This condition implies that the richer the society becomes, the more expensive it is for its members to support themselves.

Full employment in the labor force and capital
The strong assumption that labor and capital are always fully employed is represented by the conditions

$$N_i + N_s = N, \quad K_i + K_s = K. \quad (11.1.13)$$

The model synthesizes the Solow-Swan-Uzawa growth model and the Alonso location model on the basis of population growth theory. For instance, if we neglect population growth and locational aspects in our model, i.e., if $n = 0$, $v_0 = 0$ and $\eta = 0$, then our economic system is similar to the Solow-Swan-Uzawa growth model. If we neglect population and capital dynamics, i.e., if $c = 1$, $\delta_k = 0$ and $n = 0$, then our model is similar to the Alonso model, except that wage and interest rates are endogenously determined. Hence, the framework suggested in this chapter is quite general in the sense that some well-known models in theoretical economics can be considered as special cases of this model.

11.2 Economic Geographic Cycles in the Isolated Island Economy

Although the system developed in the preceding section appears very complicated, in the appendix we show that the dynamic system can be described by two-dimensional differential equations in terms of K and N as follows

$$\frac{dK}{dt} = q_1 K^{\alpha_i} N^{\beta_i} - \delta_k K,$$

$$\frac{dN}{dt} = nN\left[f(K,N) - \theta K^m N^{1-m}\right] \tag{11.2.1}$$

in which

$$f \equiv (cq_1)^q K^{\alpha_i q} N^{\beta_i q}, \quad q_1 \equiv \frac{\gamma + \xi}{s\gamma + \xi} m_1, \quad m_1 \equiv S_k^{\alpha_i} S_n^{\beta_i},$$

$$S_k \equiv \frac{\alpha_i (s\gamma + \xi)}{\alpha_i (s\gamma + \xi) + \alpha_s \gamma (1-s)},$$

$$S_n \equiv \frac{\beta_i (s\gamma + \xi)}{\beta_i (s\gamma + \xi) + \beta_s \gamma (1-s)}. \tag{11.2.2}$$

For any given $N(t)$ and $K(t)$ we can determine all the other variables in the system at any point of time. This means that through the dynamic properties of (11.2.1), all other variables are uniquely determined at any location and time period. It is thus sufficient to examine the properties of (11.2.1). In the appendix we also show that the following relations are held at any point in time for any given $N(t)$ and $K(t)$

11.2 Economic Geographic Cycles in the Isolated Island Economy

$$K_i = S_k K, \quad K_s = s_k K, \quad N_i = S_n N, \quad N_s = s_n N, \tag{11.2.3}$$

$$Y(K,N) = \frac{(\gamma + \xi) m_1 K^{\alpha_i} N^{\beta_i}}{s\gamma + \xi}, \tag{11.2.4}$$

$$C_i = \frac{(1-s)\xi Y}{\gamma + \xi}, \quad C_s = \frac{(1-s)\gamma Y}{(\gamma + \xi) p}, \tag{11.2.5}$$

$$R(L) = \frac{\eta s_0 y^2(L) N}{\left[cY/N - 2v_0 L^v/(2+v) \right] \pi L^2}, \tag{11.2.6}$$

$$R(\omega) = \frac{y^2(\omega) R(L)}{y^2(L)}, \quad 0 \le \omega \le L \tag{11.2.7}$$

in which

$$S_k \equiv \frac{\alpha_s \gamma (1-s)}{\alpha_i (s\gamma + \xi) + \alpha_s \gamma (1-s)}, \quad S_n \equiv \frac{\beta_s \gamma (1-s)}{\beta_i (s\gamma + \xi) + \beta_s \gamma (1-s)}. \tag{11.2.8}$$

The procedure for determining the values of the state variables is as follows: K and N by (11.2.1) $\to$ K_i, K_s, N_i and N_s by (11.2.3) $\to$ F_i and F_s by (11.1.1) $\to$ w, p and r by (11.1.3) $\to$ Y by (11.2.4) $\to$ C_i and C_s by (11.2.5) $\to$ $R(L)$ by (11.2.6) $\to$ $R(\omega)$ by (11.2.7) $\to$ $c_i(\omega)$, $c_s(\omega)$, and $L_h(\omega)$ by (11.1.7) $\to$ $U(\omega)$ by (11.1.4).

The motion of $K(t)$ and $N(t)$ is determined by (11.2.1). The other variables are determined as functions of $K(t)$ and $N(t)$ through the assumed conditions of perfect competition among the economic agents. That is, all other variables are "enslaved" by $K(t)$ and $N(t)$.

To determine the dynamic properties of the system, it is noted that (11.2.1) has the following unique long-run equilibrium:

$$K = \left[\frac{(cq_1)^q}{\theta} \left(\frac{q_1}{\delta} \right)^{(1-m)/\beta_i - q} \right]^Q, \quad N = \left(\frac{\delta}{q_1} \right)^{1/\beta_i} K \tag{11.2.9}$$

in which $Q \equiv 1/(1-q)$.

It is easy to check that the two eigenvalues, ϕ_1 and ϕ_2, are given by

$$\phi_{1,2} = -\frac{a_1}{2} \pm \left[\frac{a_1^2}{4} - (1-q)\beta_i \delta nf\right] \qquad (11.2.10)$$

where $a_1 \equiv \delta\beta_i - (\beta_i q - 1 + m)nf$. From (11.2.10), we obtain the following proposition.

Proposition 11.2.1.
The dynamic urban system has a unique equilibrium. The system may be either stable or unstable, depending upon the parameter values of the capital and population dynamic equations. For instance, when population growth is not strongly affected by economic conditions (i.e., $\beta_i q - 1 + m \leq 0$) or is slowly adapted to equilibrium (i.e., n being small), then the system is stable.

Let the equilibrium defined in (11.2.2) be denoted by (K_0, N_0). From (11.2.10), we see that when $a_1 = 0$, the two eigenvalues are pure imaginary. One can appropriately choose combinations of parameter values such that

$$a_1 = 0, \quad \text{i.e.,} \quad \delta\beta_i - (\beta_i q - 1 + m)nf = 0 \qquad (11.2.11)$$

which defines a critical point. Let n_0 be the value of n for which the term is equal to zero. As n is the adjustment speed of the population, the requirement is meaningful if $\beta_i q - 1 + m > 0$, which implies that the population dynamics are strongly affected by economic conditions. As f is independent of n, we have

$$n_0 = \frac{\delta\beta_i}{(\beta_i q - 1 + m)f} > 0.$$

In what follows, we denote small perturbations of n from n_0 by ε, i.e., $\varepsilon \equiv n - n_0$.

At $\varepsilon = 0$, we have

$$\phi_{1,2}(0) = \pm i\phi_0 \qquad (11.2.12)$$

11.2 Economic Geographic Cycles in the Isolated Island Economy

where $\phi_0 \equiv [(1-q)\delta\beta_i nf]^{1/2}$. That is, at $\varepsilon = 0$ one has a pair of pure imaginary eigenvalues. Let $\phi(\varepsilon)$ denote the eigenvalue which equals $i\phi_0$ at $\varepsilon = 0$. As the eigenvalues are continuous functions of ε, the function $\phi(\varepsilon)$ is well defined in the neighborhood of $\varepsilon = 0$. As

$$\phi_\varepsilon(0) = \frac{(\beta_i q - 1 + m)f}{2} - \frac{(1-q)\beta_i \delta fi}{2\phi_0} \qquad (11.2.13)$$

the real part of the derivative of the eigenvalue ϕ is positive when ε becomes positive. Hence, when ε crosses its critical value, the equilibrium becomes unstable. An increase in the adjustment speed of population may result in instabilities.

According to the Hopf bifurcation theorem (e.g., Zhang, 1991, Iooss and Joseph, 1980), we know that when ε becomes positive, cycles appear around the stationary state. The results can be summarized in the following proposition.

Proposition 11.2.2.
Near the critical state, when the adjustment speed of the population dynamics is increased, the system exhibits endogenous permanent oscillations. The cycles are approximately given by

$$K(t,\tau) = K_0 + 2\tau\delta\beta_i K_0 \cos(\phi_0 t) + O(\tau^2),$$
$$N(t,\tau) = N_0 + 2\tau N_0[\delta\beta_i \cos(\phi_0 t) - \phi_0 \sin(\phi_0 t)] + O(\tau^2) \qquad (11.2.14)$$

where τ is a small expansion parameter.

The proof of the above proposition is given in Zhang (1994b). The same reference also provides an economic interpretation. The theorem can be proved by applying the bifurcation method of Iooss and Joseph (1980). This method has been applied to some economic systems in Zhang (1991d). What should be noted, however, is that the eigenvectors X and the adjoint eigenvectors X^* are given by

$$X = (\beta_i \delta K_0, (\beta_i \delta + i\phi_0) N_0)^T,$$
$$X^* = \left(\frac{\alpha_i q - m}{K_0} nfx^*, \frac{\beta_i \delta - i\phi_0}{N_0} x^*\right)^T$$

where

$$x^* = \frac{1}{(\alpha_i q - m - 1 + q)\delta\beta_i nf + \delta^2\beta_i^2 - i2\delta\beta_i\phi_0}.$$

Moreover, the stability conditions and more accurate expressions of the limit cycles can be described using the eigenvectors and the adjoint eigenvectors (Zhang, 1991d). Since these expressions are too complicated to deepen our insight into the problem, we shall not calculate them here. We illustrate the cycle given by the above proposition as in Fig. 11.2.1.

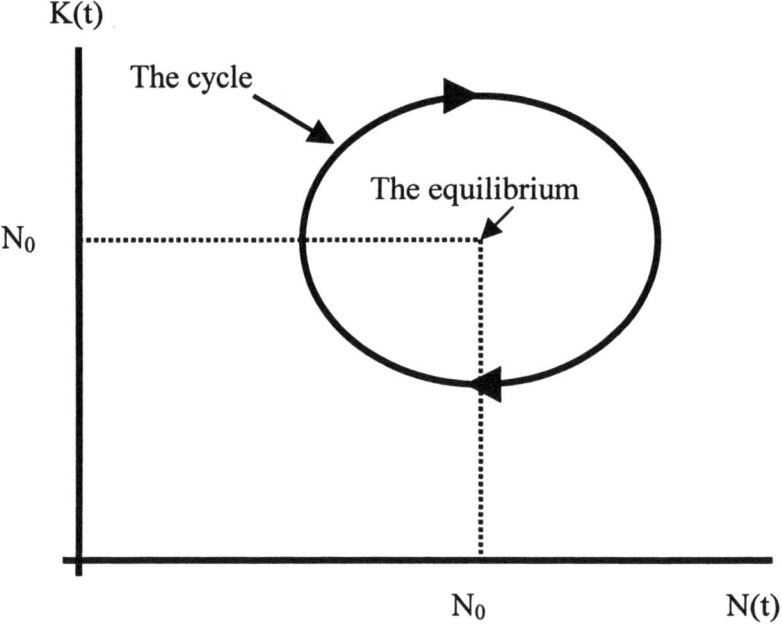

Fig.11.2.1. The Urban Cycle with the Population and Capital

As K and N exhibit oscillations, the other variables are periodic too. Limit cycles are due to instabilities of the dynamic system. It is important to note that the rent density, residential distribution and the city's boundary permanently oscillate near the stationary state given by $\varepsilon = 0$. Fig. 11.2.2 illustrates a dynamics of the land rent $R(0,t)$ near the CBD.

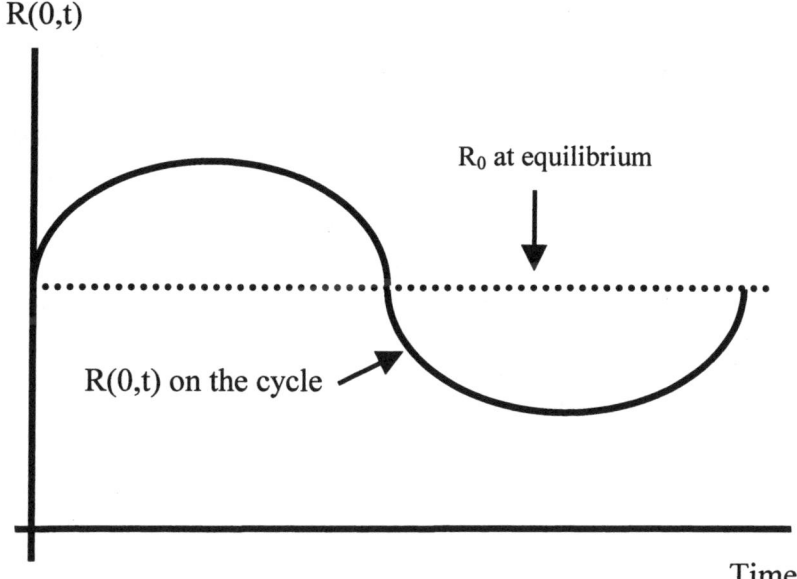

Fig. 11.2.2. The Periodic Land Rent at the CBD

11.3 Aperiodic Oscillations in the J-Island Economy

Although it is important to identify conditions for possible existence of a permanently time-dependent behavior of economic systems, the task is not easy. In fact, even in the literature of nonlinear dynamic theory only a few special nonlinear dynamic (deterministic) systems which exhibit permanent oscillations and chaos have been investigated. It is analytically difficult to identify conditions for existence of complicated time-dependent behavior in general terms. In this section, by applying analytical results from coupled oscillator dynamics, one can identify aperiodic solutions in the economic geography.

In recent years, many efforts have been undertaken to study coupled oscillator dynamics. Turing's seminar work (Turing, 1952) on morphogenesis has motivated studies of oscillators in an attempt to understand the existence of many different possible patterns of dynamic behavior (e.g., Murry, 1989, Ashwin and Swift, 1992). Various forms of mathematical models, employing the so-called coupled oscillators, have been suggested and their behavior examined with various methods in nonlinear dynamic theory. For instance, by considering the symmetries of the network and using the theories of groups and singularities, Golubitsky and Stewart (1985) identify several possible complicated nonlinear behaviors in a model for rings of cells coupled together near a Hopf bifurcation. Swift (1988) has examined the case of four cells and has found scenarios where tori branch from the quiescent state. Ashwin and Swift (1992) discuss how the limit cycles lose and gain stability, and how symmetry can

196   11 Nonlinear Dynamics of a Multi-City System

give rise to structurally stable heteroclinic cycles within a framework of arbitrary networks of identical (weakly coupled) dissipative oscillators. Lorenz (1987) provides the first example of applying the idea to identifying possible existence of chaos in the international trade model. Zhang (1991d) also applies the method to economic dynamics under returns-to-scale economies, to locate conditions for existence of aperiodic oscillations. The purpose of this section is to provide another example of possible existence of permanent oscillations in economic geographic dynamics, on the basis of the analytical results from the preceding sections.

As a pedagogical example, an economic system consisting of J islands is considered, indexed by $j = 1, ..., J$. Each island is described by the economic system defined as follows. When there are no social and economic interactions between any two islands, the dynamics of each island's economic geography are determined by the following two-dimensional system

$$\frac{dK_j}{dt} = F_j(K_j, N_j),$$
$$\frac{dN_j}{dt} = \Omega_j(K_j, N_j), \quad j = 1, ..., J \tag{11.3.1}$$

where the functions, F_j and Ω_j, are functions of $K_j(t)$ and $N_j(t)$ explicitly defined in (11.2.1). At each point in time, each island's economic geography is described by the 19 endogenous space-time-dependent variables defined in Section 11.1. In the remainder of this section, it is assumed that the parameters satisfy the requirements in Proposition 11.2.2. That is, each island's economy is oscillatory before interactions occur.

It is reasonable to assume existence of interactions among the J islands. For instance, if free movement of capital is allowed, there will be capital flows due to differences in prices and interest rates among the J economies. If people are entitled to free migration, there may exist migration among the economies as people may obtain different utility levels at various island economies. There may be many social and economic factors that determine the dynamics of capital and population. We may propose various economic mechanisms to connect the J-island economic system. For instance, in the case that we neglect possible differences in culture and human quality, free migration will result in an identity among utility levels at the J islands. Even if we introduce differences in people's preferences, a household will choose the island which maximizes its utility level. We may similarly discuss movement of capital and goods.

Let us generally assume that the dynamics of the J-island economies can be described by the following system

11.3 Aperiodic Oscillations in the J-Island Economy

$$\frac{dK_j}{dt} = F_j^*(K, N; \mu),$$

$$\frac{dN_j}{dt} = \Omega_j^*(K, N; \mu), \quad j = 1, ..., J \tag{11.3.2}$$

where

$$K \equiv (K_1, K_2, ..., K_J), \quad N \equiv (N_1, N_1, ..., N_J)$$

are capital and population vectors, respectively. In (11.3.1), the functional forms, $F_j^*(K, N; \mu)$ and $\Omega_j^*(K, N; \mu)$, of the capital and population dynamics in the trade system are determined by specified institutional and economic structures under consideration. We call μ (≥ 0) the trade degree parameter. When $\mu = 0$, there is no trade. That is, the functions, $F_j^*(K, N; \mu)$ and $\Omega_j^*(K, N; \mu)$, have to satisfy the following requirement

$$F_j^*(K, N; 0) = F_j(K, N), \quad \Omega_j^*(K, N; 0) = \Omega_j(K, N). \tag{11.3.3}$$

From the literature mentioned at the beginning of this section, we know that an accurate analysis of even a simple case of the above system is very complicated. Therefore, one can only provide some general ideas, instead of giving an accurate specification of the functional forms and strict analysis.

The concern is with what will happen to the system if μ is positive. Obviously the extended system (11.3.2) is now made up of J coupled limit oscillators. Introduce

$$x = (K_1, N_1, K_2, N_2, ..., K_J, N_J).$$

For $\mu = 0$, x is a quasiperiodic solution of (17.3.2). We may write the quasiperiodic solution $x(t)$ in the form of $x(t) = X(v_1 t, v_2 t, ... v_J t)$ where X is periodic 1 separately in its J arguments, and the frequencies v_j ($j = 1, 2, ..., J$) are rationally independent. As demonstrated by Newhouse, Ruelle and Takens (1978) and Ruelle (1989), a perturbation of a quasiperiodic flow on T^J for $J \geq 3$ leads to complicated behavior. Quasiperiodicity with J independent frequencies may persist, or frequency locking may lead to periodic orbits or to quasiperiodic motion with less than J independent frequencies. Furthermore, chaos may also appear. Strange attractors arise by suitable arbitrary small perturbations of any given quasiperiodic flow. It should be noted that if we assume

11.4 On Spatial Chaos

This chapter developed a complicated dynamics of urban systems. From the previous chapters, we might see many possible ways to extend the model in this chapter. In addition to the economic interactions which neoclassical growth theory and neoclassical urban economics attempt to explain, our model may also shed some light upon the urban dynamic phenomena which the "path-dependent processes of the spatial location" approach (e.g., Authur, 1989, 1990, Dendrinos and Sonis, 1990, Rosser, 1991, and Weidlich and Haag, 1983) tries to explore. This new approach of explaining the rise and fall of cities and regions is much affected by modern nonlinear dynamic theory (e.g., Zhang, 1991c). As the economic structure in our model may have multiple equilibria and the system may be unstable, it is reasonable to expect that the actual urban dynamics are path-dependent.

Appendix

A.11.1 The Dynamics of Capital and the Population

We now prove (11.2.1). First, we notice that the utility level among households is equal over space, i.e., $U(\omega,t) = U(L,T)$, for all $0 \leq \omega \leq L$. From (11.1.4) and (11.1.7), we obtain (11.2.7). Since $y(\omega_1) \geq y(\omega_2)$ for any $0 \leq \omega_1 \leq \omega_2 \leq L$, from (11.2.7) we see that the land rent function monotonically decreases, i.e.

$$\frac{dR(\omega,t)}{d\omega} < 0, \quad 0 \leq \omega \leq L.$$

From $n(\omega) = 1/L_h(\omega)$ and (11.1.7), we get

$$n(\omega) = \frac{R(\omega)}{\eta s_0 y(\omega)}. \tag{11.A.1.1}$$

Substituting (11.2.7) into (11.A.1.1) yields

$$n(\omega) = \frac{y(\omega)R(L)}{\eta s_0 y(L)^2}. \tag{11.A.1.2}$$

Using the above equation and (11.1.9), we have

$$R(L) = \frac{\eta s_0 y(L)^2 N}{\{cY/N - 2\upsilon_0 L^\upsilon/(2+\upsilon)\}\pi L^2}. \tag{11.A.1.3}$$

The above equation determines the land rent at the perimeter of the island for given Y and N.

Substituting (11.1.7) and (11.2.7) into (2.1.10) and then calculating the integration, we can express C_i and C_s as functions of $N(t)$ and $Y(t)$ as follows

$$C_i = \xi G^*(N,Y), \quad C_s = \frac{\gamma G^*(N,Y)}{p(t)} \tag{11.A.1.4}$$

where

$$G^*(N,Y) \equiv \frac{\pi}{\eta} \int_0^L \omega R(\omega) d\omega. \tag{11.A.1.5}$$

From

$$C_s = F_s, \quad C_i + sY, \quad Y = F_i + pF_s$$

and (11.A.1.5), we get

$$G^*(Y) = \frac{(1-s)Y}{\gamma + \xi}.$$

Substituting this equation into (11.A.1.4) yields

$$C_s = \frac{(1-s)\gamma Y}{\gamma + \xi}, \quad C_s = \frac{(1-s)\xi Y}{(\gamma + \xi)p}. \tag{11.A.1.6}$$

From $r = \alpha_i F_i / K_i = \alpha_s pF_s / K_s$, $C_s = F_s$ and (11.A.1.6), we get

$$\frac{K_s}{K_i} = \frac{(1-s)\gamma Y \alpha_s Y}{(\gamma + \xi)\alpha_i F_i}.$$ (11.A.1.7)

Substituting

$$F_i = C_i + sY = \left\{\frac{(1-s)\xi}{\gamma + \xi} + s\right\}Y$$ (11.A.1.8)

into (11.A.1.7) yields

$$\frac{K_s}{K_i} = \frac{(1-s)\gamma \alpha_s}{(s\gamma + \xi)\alpha_i}.$$ (11.A.1.9)

From (11.1.3), we obtain

$$\frac{K_s}{K_i} = \frac{\alpha_s \beta_i N_s}{\alpha_i \beta_s N_i}.$$ (11.A.1.10)

From (11.A.1.9) and (11.A.1.10), we solve

$$\frac{N_s}{N_i} = \frac{(1-s)\gamma \beta_s}{(s\gamma + \xi)\beta_i}.$$ (11.A.1.11)

From (11.A.1.9), (11.A.1.11) and (11.1.13), we get (11.1.12). Substituting (11.1.12) into (11.1.1), we solve F_i and F_s as functions of N and K as follows

$$F_i(K,N) = m_1 K^{\alpha_i} N^{\beta_i}, \quad F_s(K,N) = m_2 K^{\alpha_s} N^{\beta_s}$$ (11.A.1.12)

in which $m_1 \equiv S_k^{\alpha_i} S_n^{\beta_i}$ and $m_2 \equiv s_k^{\alpha_s} s_n^{\beta_s}$. By (11.A.1.12) and (11.A.1.8), we get (11.2.4). From (11.1.11), (11.1.12), (11.A.1.12) and (11.A.1.8), we get (11.2.1).

12 Further Issues on Cities

The soil for the edifice must be explored by critique as deep down as the foundation of the faculty of principles independent of experience, in order that it may sink in no part, for this would inevitably bring about the downfall of the whole.

Kant

This book applied the basic concepts and frameworks on growth theory (without space) and trade theory by Zhang (1999, 2000) to explore complex interdependence between economic growth and spatial pattern formation with endogenous capital and knowledge. Since our main purpose was to construct a theoretical framework to synthesize urban economics and neoclassical growth theory in a consistent way, we were concerned only with simplified structures of spatial economies. However, it is not difficult to extend the models in this book to examine other situations of spatial economies. For instance, as in Beckmann (1968), we may distinguish four different locational patterns according to the space requirements of producers and the population density of consumers: (1) producers and consumers are concentrated at point locations and in point markets; (2) producers are located in points, consumers extended through the market area; (3) consumers are concentrated at points, producers dispersed; and (4) both producers and consumers are extended through the market area. We may treat space in spatially continuous frameworks (Beckmann and Puu, 1985, Andersson and Zhang, 1988, Zhang, 1988a, 1988b, 1989b, 1989c). This book explored the complexity of spatial economies within a perfectly competitive framework. Our focus on perfect competition with endogenous capital and knowledge is different from the contemporary mainstream of spatial economics that is mainly concerned with monopolistic competition or fixed prices frameworks (e.g., Arnott, 1996a, 1996b, Greenhut, Norman, and Hung, 1987).

We tried to provide a systematic treatment of spatial economies in a deterministic dynamic context. We developed dynamic spatial models to deal with the issues related to the relationship between knowledge, growth, and spatial economic structures with heterogeneous population. It is not intended to be a comprehensive treatment of all the important issues related to spatial economies. Since the models proposed in this book were consistent with each other, it is conceptually easy to make further extensions of our models. We now mention a few directions of possible extensions of our work.

Scale effects of endogenous population
From the literature of classical economics we know at least four input factors which may exhibit increasing or decreasing returns to scale effects in economic dynamics: environment and infrastructures (of transportation and communication systems), institutions, knowledge, and population, (e.g., Malthus, 1933, Marshall, 1890, Haavelmo, 1954, Niehans, 1963, Zhang, 1999, 2000). Although we introduced endogenous population into our analytical framework, we did not take account of age structures of the population.

Networks and infrastructures
In a long-term analysis, it is necessary to examine decision-making processes involved in construction and maintenance of infrastructures (e.g., Takayama and Judge, 1971, Andersson et al., 1990, Johansson, 1995, Batten, 1982, Mun, 1997). Channels, roads, railways and airline systems determine the mobility and the costs associated with movements of people and goods. They are essential for understanding spatial economies.

Knowledge and human capital
We did not deal with economic evolution with knowledge in a comprehensive way because of analytical difficulties for obtaining explicit conclusions. Following the modeling framework developed by Zhang (1999, 2000), it is conceptually possible to introduce other aspects such as human capital structures and professional amenities into our framework.

Multiple kinds of capital, people and natural resources
It is not difficult to relax the assumption of a single kind of capital (Burmeister and Dobell, 1970, Barro and Sala-i-Martin, 1995). The introduction of multiple capital goods will cause analytical difficulties. It should be noted that the traditional neoclassical growth theory did not succeed in dealing with growth issues with multiple capital goods in the sense that the consumer behavior was not properly modeled (Zhang, 1999). We developed multi-group trade models, but our classification of labor force was simplified. Different kinds of labor force may enter production systems in different ways.

Preference structures and government policy
Utility functions may be taken on various forms. Except common issues related to forms of utility functions (Lancaster, 1966, 1971), it seems important to further examine dynamics of preference changes and spatial economic development. It should be remarked that Dixit and Stiglitz (1977) emphasized the implications of the endogenization of the number of goods for productivity progress and R&D. This idea can be taken into account by assuming that knowledge affects the parameters in the utility and production functions.

Governments may intervene economic systems in different ways within competitive economic systems. It is important to examine the impact of various government interventions on spatial economies.

Further Issues on Cities

Full employment of production factors and monetary variables

We assumed that production factors such as labor force, capital, and land are always fully employed. These assumptions should be relaxed in order to analyze modern economies. Issues related unemployment and economic evolution can be analyzed within the framework suggested in Zhang (1999).

We assume that spatial economic exchanges take place in the form of barter; in other words, money is treated as a veil, which has no impact on the underlying variables but serves as a reference unit, the *numeraire* (Friedman and Hahn, 1990). This omission of monetary aspects does not mean that we consider it unimportant to integrate monetary economics with our approach. We assumed that monetary variables are fast in the sense that their values are determined by their marginal values at any point of time. In reality, monetary variables may seldom be adjusted so quickly. We may generally denote monetary dynamics in the following general form

$$\frac{dp}{dt} = \varepsilon G(p, K)$$

where p is a vector of monetary variables such as money, exchange rates and prices of goods and ε is the adjustment speed vector. Different theories can be applied to specify functional forms of $G(p, K)$.

Bibliography

Abdel-Rahman, H.M. (1988): Product Differentiation, Monopolistic Competition and City Size. Regional Science and Urban Economics 18, 69-86

Aghion, P., Howitt, P. (1992): A Model of Growth through Creative Destruction. Econometrica 60, 323-351

Aghion, P., Howitt, P. (1998): Endogenous Growth Theory. Mass., Cambridge: The MIT Press

Allen, P., Sanglier, M. (1979): A Dynamic Model of a Central Place System. Geographical Analysis 11, 256-272

Alonso, W. (1964): Location and Land Use. MA., Cambridge: Harvard University Press

Alperovich, G. (1992): Economic Development and Population Concentration. Economic Development and Cultural Change 41, 63-74

Anas, A. (1982): Dynamics of Urban Residential Growth. Journal of Urban Economics 5, 66-87

Anas, A. (1992): On the Birth and Growth of Cities - Laissez-Faire and Planned Compared. Regional Science and Urban Economics 22, 243-258

Anderson, P.W., Arrow, J.K., Pines, D. (1988, Eds.): The Economy as an Evolving Complex System. New York: Addison-Wesley

Andersson, Å.E. (1986): The Four Logistical Revolutions. Papers of Regional Science Association 59, 1-12

Andersson, Å.E., Anderstig, C., Hårsman, B. (1990): Knowledge and Communications Infrastructure and Regional Economic Change. Regional Science and Urban Economics 20, 359-376

Andersson, Å.E., Zhang, W.B. (1988): The Two Dimensional Continuous Spatial Input-Output System. Ricerche Economiche XLII, 2, 222-242

Andersson, Å.E., Zhang, W.B. (1990): Endogenous Technological Changes and Economic Growth. In Chatterji, M., Kuenne, R. (Eds.): Dynamics and Conflict in Regional Structural Change. London: The Macmillan Press LTD

Arnott, R. (1996a, Ed.): Regional and Urban Economics, Part 1. Amsterdam: Harwood Academic Publishers

Arnott, R. (1996b, Ed.): Regional and Urban Economics, Part 2. Amsterdam: Harwood Academic Publishers

Arnott, R.J. (1979): Unpriced Transport Congestion. Journal of Economic Theory 21, 294-316

Arnott, R. (1987): Economic Theory and Housing, in Handbook of Regional and Urban Economics, Volume II, edited by E.S. Mills. Amsterdam: North-Holland

Arrow, K.J. (1962): The Economic Implications of Learning by Doing. Review of Economic Studies 29: 155-173

Arthur, W.B. (1989): Competing Technologies, Increasing Returns, and Lock-in by Historical Events. Economic Journal 99, 116-131

Arthur, W.B. (1990): 'Silicon Valley' Locational Clusters When Do Increasing Returns Imply Monopoly?. Mathematical Social Sciences 19, 235-251

Asami, Y., Fujita, M., Smith, T.E. (1990): On the Foundations of Land Use Theory - Discrete versus Continuous Populations. Regional Science and Urban Economics 20, 473-508

Ashwin, P., Swift, J.W. (1992): The Dynamics of n Weakly Coupled Identical Oscillators. Journal of Nonlinear Science 2, 69-108

Auer, L.von. (1998): Dynamic Preferences, Choice Mechanisms, and Welfare. Berlin: Spriner

Bardhan, P.K. (1965): Optimum Capital Accumulation and International Trade. Review of Economic Studies 32, 241-244.

Barro, R.J., Sala-i-Martin, X. (1995): Economic Growth. New York: McGraw-Hill, Inc

Batten, D.F. (1982): The Interregional Linkages Between National and Regional Input-Output Analysis. International Regional Science Review 7, 53-68

Becker, G.S. (1965): A Theory of the Allocation of Time. Economic Journal 75, 493-517

Becker, G.S. (1976): The Economic Approach to Human Behavior. Chicago: The University of Chicago Press

Becker, G.S. (1981): A Treatise on the Family. Mass., Cambridge: Harvard University Press

Becker, G.S. (1985): Human Capital, Effort, and the Sexy Division of Labor. Journal of Labor Economics 3, S34-58

Beckmann, M.J. (1957): On the Distribution of Rent and Residential Density in Cities, Paper presented at the inter-Departmental Seminar on Mathematical Applications in the Social Sciences, Yale University

Beckmann, M.J. (1968): Location Theory. New York: Harper & Brothers.

Beckmann, M.J. (1969): On the Distribution of Urban Rent and Residential Density. Journal of Economic Theory 1, 60-68

Beckmann, M.J. (1973): Equilibrium Models of Residential Land Use. Regional Science and Urban Economics 3, 361-368

Beckmann, M.J. (1976): Spatial Equilibrium in a Dispersed City, in Mathematical Land Use Theory, edited by Y.Y. Papageorgiou. Lexington: Lexington Books

Beckmann, M.J. (1987): Location of Economic Activity, in The New Palgrave: A Dictionary of Economics 3, edited by J. Eatwell, M. Milgate, and P. Newman. New York: Stockton Press

Beckmann, M.J. (1990): Taste and Welfare in Travellers' Choices. Studies in Regional Science 20, No-2, 25-29

Beckmann, M., Puu, T. (1985): Spatial Economics: Density, Potential, and Flow. Amsterdam: North-Holland

Bell, C. (1991): Regional Heterogeneity, Migration, and Shadow Prices. Journal of Public Economics 46, 1-27

Benson, B. (1980): Löschian Competition Under Alternative Demand Conditions. American Economic Review 70, 1098-105

Benson, B. (1984): Spatial Competition with Free Entry, Chamberlinian Tangencies, and Social Efficiency. Journal of Urban Economics 15, 270-86

Berliant, M., Ten Raa, T. (1987): On the Continuum Approach of Spatial and Some Local Public Goods or Product Differentiation Models. Working Paper 72, Rochester Center for Economic Research, University of Rochester

Blaug, M. (1985): Economic Theory in Retrospect, fourth edition. Cambridge: Cambridge University Press

Blomquist, G.C., Berger, M.C., Hoehn, J.C. (1988): New Estimates of Quality of Life in Urban Areas. American Economic Review 78, 89-107

Boyer, M. (1978): A Habit Forming Optimal Growth Model. International Economic Review 19, 585-609

Brueckner, J.K. (1981): A Dynamic Model of Housing Production. Journal of Urban Economics 10, 1-14

Brueckner, J.K., Rabenau, B.V. (1981): Dynamics of Land Use for a Closed City. Regional Science and Urban Economics 11, 1-17

Buiter, W.H. (1981): Time Preference and International Lending and Borrowing in an Overlapping Generations Model. Journal of Political Economy 89, 769-797

Burmeister, E., Dobell, A.R. (1970): Mathematical Theories of Economic Growth. London: Collier Macmillan Publishers

Calem, P.S., Carlino, G.A. (1991): Urban Agglomeration Economies in the Presence of Technical Change. Journal of Urban Economics 29, 82-95

Cheshire, P.C., Evans, A.W. (1991, Eds.): Urban and Regional Economics. Hants: Edward Elgar Publishing Company

Chiappori, P.A. (1988): Rational Household Labor Supply. Econometrica 51, 63-89

Chiappori, P.A. (1992): Collective Labor Supply and Welfare. Journal of Political Economy 100, 438-467

Christaller, W. (1933): Central Places in Southern Germany, translated from the German origin by C.W., Baskin, 1966. N.J., Englewood Cliffs: Prentice-Hall

Cole, H.L., Mailath, G.J., Postlewaite, A. (1992): Social Norms, Savings Behavior, and Growth. Journal of Political Economy 100, 1092-125

David, P.A., Rosenbloom, J.L. (1990): Marshallian Factor Market Externalities and the Dynamics of Industrial Localization. Journal of Urban Economics 28, 349-370

Dendrinos, D.S. (1992): The Dynamics of Cities, Ecological Determinism, Dualism and Chaos. London: Routledge

Dendrinos, D.S., Sonis, M. (1990): Chaos and Socio-Spatial Dynamics. New York: Springer-Verlag

Devereux, M.B., Shi, S. (1991): Capital Accumulation and the Current Account in a Two-Country Model. Journal of International Economics 30, 1-25

D'Aspremont, C., Gabszewicz, J.J., Thisse, J.F. (1979): On Hotelling's Stability in Competition. Econometrica 47, 1145-1150

Diamond, D.B., Tolley, G.S. (1981, Eds.): The Economics of Urban Amenities. New York: Academic Press

Dierx, A.H. (1990): Intermetropolitan Transfer Costs and the Equilibrium Size of Metropolitan Areas. Regional Science and Urban Economics 20, 173-187.

Dixit, A. (1973): The Optimum Factory Town. Bell J. Econom. Management Sci. 4, 637-51

Dixit, A.K., Stiglitz, J.E. (1977): Monopolistic Competition and Optimum Product Diversity. American Economic Review 67, 297-308

Dixon, H.D., Rankin, N. (1994): Imperfect Competition and Macroeconomics: A Survey. Oxford Economic Papers 46, 171-199

Dollar, D. (1986): Technological Innovation, Capital Mobility, and the Product Cycle in the North-South Trade. American Economic Review 76, 177-190

Dosi, G., Pavitt, K., Soete, L. (1990): The Economics of Technical Change and International Trade. New York: Harvester Whertsheaf

Drandakis, E., Phelps, E.S. (1966): A Model of Induced Invention, Growth and Distribution. Economic Journal LXXVI, 823-40

Eaton, B.C., Lipsey, R.G. (1975): The Principle of Minimum Differentiation Reconsidered: Some New Developments in the Theory of Spatial Competition. Review of Economic Studies 42: 27-50

Eaton, B.C., Lipsey, R.G. (1975): The Nonuniqueness of Equilibrium in the Löschian Model. American Economic Review 66, 77-93

Ethier, W.J. (1982): National and International Returns to Scale in the Modern Theory of International Trade. American Economic Review 72, 389-405

Fershtman, C., Weiss, Y. (1993): Social Status, Culture and Economic Performance. The Economic Journal 103, 946-959

Findlay, R., Wellisz, S. (1993, Eds.): Five Small Open Economies. New York: Oxford University Press

Fisher, I. (1930): The Theory of Interest. New York: Macmillan.

Frenkel, J., Razin, A. (1987): Fiscal Policy and the World Economy. MA., Cambridge: MIT Press

Friedman, B.M., Hahn, F.H. (1990, Eds.): Handbook of Monetary Economics, Volumes I, II. Amsterdam: North-Holland

Fujita, M. (1982): Spatial Patterns of Residential Development. Journal of Urban Economic 12, 22-52

Fujita, M. (1989): Urban Economic Theory - Land Use and City Size. Cambridge: Cambridge University Press

Fujita, M., Krugman, P., Venables, A.J. (1999): The Spatial Economy – Cities, Regions, and International Trade. Mass., Cambridge: The MIT Press

Fung, K.M., Ishikawa, J. (1992): Dynamic Increasing Returns, Technology and Economic Growth in a Small Open Economy. Journal of Development Economics 35, 63-87

Gersovitz, M. (1988): Saving and Development. In Chenery, H., Srinivasan, T.N. (Eds.): Handbook of Development Economics, Volume I. Amsterdam: North Holland

Golubitsky, M., Steward, I.N. (1985): Hopf Bifurcation with Dihedral Group Symmetry in Multiparameter Bifurcation Theory. Contemporary Mathematics 56, 131-173, AMS, Providence, RI

Greenhut, J., Greenhut, M.L. (1975): Spatial Price Discrimination, Competition and Locational Effects. Economica 42, 401-419

Greenhut, M.L. (1956): Plant Location in Theory and Practice: The Economics of Space. Chapel Hill: University of North Carolina Press

Greenhut, M.L. (1963): Microeconomics and the Space Economy. Chicago: University of Chicago Press

Greenhut, M.L., Ohta, H. (1973): Monopoly Output Under Alternative Spatial Pricing Techniques. American Economic Review 62, 705-13

Greenhut, M.L., Ohta, H. (1975): A Theoretical Mapping From Perfect Competition to Imperfect Competition. Southern Economic Journal 42, 177-92

Greenhut, M.L., Norman, G., Hung, C.S. (1987): The Economics of Imperfect Competition - A Spatial Approach. Cambridge: Cambridge University Press

Grossman, G.M., Helpman, E. (1991): Innovation and Growth in the Global Economy. Cambridge, Mass.: The MIT Press

Haavelmo, T. (1954): A Study in the Theory of Economic Evolution. Amsterdam: North-Holland

Haken, H. (1977): Synergetics: An Introduction. Berlin: Springer-Verlag

Haken, H. (1983): Advanced Synergetics - Instability Hierarchies of Self-Organizing Systems and Devices. Berlin: Springer

Hartwick, P.G., Hartwick, J.M. (1974): Efficient Resource Allocation in a Multinucleated City with Intermediate Goods. Quarterly Journal of Economics 88, 340-52

Heckman, J.J., Macurdy, T.E. (1980): A Life Cycle Model of Labor Supply. Review of Economic Studies 47, 47-74

Helsley, R.W., Starnge, W.C. (1991): Agglomeration Economies and Urban Capital Markets. Journal of Urban Economics 29, 96-112

Helsley, R.W., Sullivan, A.M. (1991): Urban Subcenter Formation. Regional Science and Urban Economics 21, 255-76

Henderson, J.V. (1974): The Sizes and Types of Cities. American Economic Review 64, 640-656

Henderson, J.V. (1985): Economic Theories and Cities. 2nd edition, Academic Press, New York

Henderson, J.V. (1986): The Efficiency of Resource Usage and City Size. Journal of Urban Economics 19, 47-70

Herbert, D.J., Stevens, B.H. (1960): A Model for the Distribution of Residential Activity in Urban Areas. Journal of Regional Science 2, 21-36

Hockman, O., Pines, D. (1980a): Costs Adjustment and Demolition Costs in Residential Construction and Their Effects on Urban Growth. Journal of Urban Economics 7, 2-19

Hockman, O., Pines, D. (1980b): Costs of Adjustment and the Spatial Pattern of a Growing Open City. Econometrica 50, 1371-1389

Hoover, E.M. Jr. (1937a): Spatial Price Discrimination. Review of Economic Studies 4, 182-91

Hoover, E.M. Jr. (1937b): Location Theory and the Shoe and Leather Industries. Cambridge: Harvard University Press

Hotelling, H. (1929): Stability in Competition. Economic Journal 39: 41-57

Ikeda, S., Ono, Y. (1992): Macroeconomic Dynamics in a Multi-Country Economy - A Dynamic Optimization Approach. International Economic Review 33, 629-644

Iooss, G., Joseph, D.D. (1980): Elementary Stability and Bifurcation Theory. New York: Springer-Verlag

Isard, W. (1953): Some Empirical Results and Problems ofRegional Input-Output Analysis, in W. Leontief, et al., Studies in the Structure of the American Economy. New York: Oxford University Press

Isard, W. (1956): Location and Space Economy. MA., Cambridge: MIT Press

Isard, W. (1960): Methods of Regional Analysis. Mass., Cambridge: MIT Press

Johansson, B. (1995): The Dynamics of Economic Networks, in Networks in Action, edited by D. Batten, J. Casti, and R. Thord. Springer-Verlag, Berlin

Johansson, B., Karlsson, C. (1990, Eds.): Innovation, Industrial Knowledge and Trade - A Nordic Perspective. Umeå: CERUM at University of Umeå

Jones, R.W. (1971): A Three-Factor Model in Theory, Trade and History in Trade, Balance of Payments and Growth edited by J. Bhagwati, et al. Amsterdam: North-Holland

Kanemoto, Y. (1980): Theories of Urban Externalities. Amsterdam: North-Holland.

Kanemoto, Y. (1996): Externalities in Space. In Arnott, R. (1996a, Ed.): Regional and Urban Economics, Part 1. Amsterdam: Harwood Academic Publishers

Karlquist, A., Lundqvist, L., Snickars, F. (1975, Eds.): Dynamic Allocation of Urban Space. Lexington: Lexington Books

Kennedy, C. (1964): Induced Bias in Innovation and the Theory of Distribution. Economic Journal LXXIV, September

Kern, C.R. (1981): Racial Prejudice and Residential Segregation: The Yinger Model Revisited. Journal of Urban Economics 10, 164-72

Kobayashi, K., Zhang, W.B., Yoshikawa, K. (1986): Theoretical Analysis on Taste Changes and Developer's Behavior. Infrastructure Planning, No. 4, 141-146 (in Japanese)

Kobayashi, K., Zhang, W.B., Yoshikawa, K. (1989): Taste Changes and Conservation Laws in the Housing Market. In Andersson, Å.E., Batten, D., Johansson, B., Nijkamp, P. (Eds.): Spatial Theory and Dynamics. Berlin: Springer-Verlag

Krugman, P.R. (1991): Increasing Returns and Economic Geography. Journal of Political Economy 99, 483-499

Lancaster, K. (1966): A New Approach to Consumer Theory. Journal of Political Economy 74, 132-157

Lancaster, K. (1971): Consumer Demand. New York: Columbia University Press

Lorenz, H.W. (1987): International Trade and Possible Occurrence of Chaos. Economics Letters 23, 135-138

Lorenz, H.W. (1993): Nonlinear Dynamical Economics and Chaotic Motion, 2nd edition, Berlin: Springer-Verlag

Lösch, A. (1938): The Nature of Economic Regions. Southern Economic Journal 5: 71-78

Lösch, A. (1940): The Economics of Location, translated from the German origin by W.H. Woglom, 1954. New Haven: Yale University Press

Lucas, R.E. (1988): On the Mechanics of Economic Development. Journal of Monetary Economics 22, 3-42

Malthus, T.R. (1933): An Essay on Population. London: J.D. Deut

Marshall, A. (1890): Principles of Economics, 1990. London: Macmillan
Mills, E.S. (1972): Studies in the Structure of the Urban Economy. Baltimore: The Johns Hopkins Press
Mills, E.S. (1987, Ed.): Handbook of Regional and Urban Economics - Volume II, Urban Economics. Amsterdam: North-Holland
Mills, E.S., Hamilton, B.W. (1985): Urban Economics. London: Foresman and Company
Mills, E.S. (1972): Studies in the Structure of the Urban Economy. Baltimore: The Johns Hopkins Press
Mills, E.S. (1987, Ed.): Handbook of Regional and Urban Economics - Volume II, Urban Economics. Amsterdam: North-Holland
Miyao, T. (1981): Dynamic Analysis of the Urban Economy. New York: Academic Press
Miyao, T. (1987a): Dynamic Urban Models. In Mills, E.S. (Ed.) Handbook of Regional and Urban Economics, Vol. II. Amsterdam: North-Holland
Miyao, T. (1987b): Long-run Urban Growth with Agglomeration Economies Environment and Planning A 19, 1083-92
Modigliani, F. (1986): Life Cycle, Individual Thrift and the Wealth of Nations. American Economic Review 76, 297-313
Mohring, H. (1961): Land Values and the Measurement of Highway Benefits. Journal of Political Economy 69, 236-249
Moses, L.M. (1958): Location and the Theory of Production. Quarterly Journal of Economics 72: 259-272
Mun, S.I. (1997): Transport Network and System of Cities. Journal of Urban Economics 42, 205-221
Murry, J.D. (1989): Mathematical Biology. Berlin: Springer
Muth, R.F. (1961): The Spatial Structure of the Housing Market. Papers and Proceedings of the Regional Science Association 7, 207-220
Muth, R.F. (1969): Cities and Housing. Chicago: University of Chicago Press
Muth, R.F. (1973): A Vintage Model of the Housing Stock. Papers of the Regional Science Association 30, 141-156
Nelson, R.R., Winter, S.G. (1982): An Evolutionary Theory of Economic Change. MA., Cambridge: Harvard University Press
Newhouse, R., Ruelle, D., Takens, F. (1978): Occurrence of Strange Attractors - An Axiom near Quasi-periodic Flows on T^m, $m \geq 3$. Communications in Mathematical Physics 64, 35-41
Nicolis, G., Prigogine, I. (1977): Self Organization in Nonequilibrium Systems. New York: Wiley
Niehans, J. (1963): Economic Growth with Two Endogenous Factors. Quarterly Journal of Economics 77
Nijkamp, P. (1986, Ed.): Handbook of Regional and Urban Economics - Volume 1, Regional Economics. Amsterdam: North-Holland
Nijkamp, P., Reggiani, A. (1992): Interaction, Evolution and Chaos in Space. Berlin: Springer-Verlag
Nijkamp, P., Stough, R., Verhoff, E. (1998, Eds.): Symposium "Endogenous Growth in a Regional Context." *The Annals of Regional Science* 32, No.1

Norman, G. (1979): Economies of Scale, Transport Costs and Location. London: Martinus Nijhoff Publishing
Norman, G. (1981): Spatial Competition and Spatial Price Discrimination. Review of Economic Studies 48, 97-111
Ogawa, H., Fujita, M. (1980): Equilibrium Land Use Pattern in a Non-Monocentric City. Journal of Regional Science 20, 455-475
Ohta, H. (1976) On Efficiency of Production Under Conditions of Imperfect Competition. Southern Economic Journal 43, 1124-35
Oniki, H., Uzawa, H. (1965): Patterns of Trade and Investment in a Dynamic Model of International Trade. Review of Economic Studies XXXII, 15-38
Palma, A.D., Papageorgiou, Y.Y. (1988): Heterogeneity in Tastes and Urban Structure, Regional Science and Urban Economics 18, 37-56
Panico, C., Salvadori, N. (Eds.) (1993): Post Keynesian Theory of Growth and Distribution. Vermont: Elward Elgar Publishing Limited
Papageorgiou, Y.Y., Pines, D. (1990): The Logistical Foundations of Urban Economics Are Consistent. Journal of Economic Theory 50, 37-53
Papageorgiou, Y.Y., Pines, D. (1999): An Essay on Urban Economic Theory. Boston: Kluwer Academic Publishers
Pasinetti, L.L. (1974): Growth and Income Distribution - Essays in Economic Theory. Cambridge: Cambridge University Press
Persson, I., Jonung, C. (Eds.) (1997): Economics of the Family and Family Policies. London: Routledge
Persson, I., Jonung, C. (Eds.) (1998.): Women's Work and Wages. London: Routledge
Phelps, E.S. (1966): Models of Technical Progress and the Golden Rule of Research. Review of Economic Studies XXXIII, 133-45
Phlips, L. (1983): The Economics of Price Discrimination. Cambridge: Cambridge University Press
Pitchford, J.D. (1974): Population in Economic Growth. Amsterdam: North-Holland Publishing Company
Ponsard, C. (1983): History of Spatial Economic Theory. Berlin: Springer-Verlag.
Prigogine, I. (1980): From Being to Becoming. San Francisco:W.H. Freeman
Prigogine, I. (1997): The End of Certainty - Time, Chaos, and the New Laws of Nature, written in collaboration with I. Stengers. New York: The Free Press
Prigogine, I., Stengers, I. (1984): Order out of Chaos: Man's Dialogue with Nature. Boulder: New Science Library
Puu, T. (1989): Nonlinear Economic Dynamics. Berlin: Springer-Verlag.
Rabenau, B.V. (1979): Urban Growth with Agglomeration Economics and Diseconomics. Geographia Polonica 42, 77-90
Ram, R. (1982): Dependency Rates and Aggregate Savings: A New International Cross-section Study. American Economic Review 72, 537-44
Ramsey, F. (1928): A Mathematical Theory of Saving. Economic Journal 38, 543-559
Rauch, J.E. (1991): Comparative Advantage, Geographic Advantage and the Volume of Trade. The Economic Journal 101, 1230-1244
Rescher, N. (1998): Complexity – A Philosophical Overview. London: Transaction Publishers

Richardson, B.V. (1973): Regional Growth Theory. New York: John Wiley
Richardson, H.W. (1977): Regional Growth Theory. London: Macmillan
Rivera-Batiz, F. (1988): Increasing Returns, Monopolistic Competition, and Agglomeration Economies in Competition and Production. Regional Science and Urban Economics 18, 125-53
Roback, J. (1982): Wages, Rents and Quality of Life. Journal of Political Economy 90, 1257-1278
Robson, A.J. (1980): Costly Innovation and Natural Resources. International Economic Review 21, 17-30
Rodriguez, C.A. (1975): Brain Drain and Economic Growth - a Dynamic Model. Journal of Development Economics 2, 223-247
Romer, P.M. (1986): Increasing Returns and Long-Run Growth. Journal of Political Economy 94, 1002-1037
Romer, P.M. (1987): Growth Based on Increasing Returns Due to Specialization. American Economic Review 77, 56-62
Rose-Ackerman, S. (1977): The Political Economy of a Racist Housing Market. Journal of Urban Economics 4, 150-69
Rosser, J.B.Jr. (1991): From Catastrophe to Chaos: A General Theory of Economic Discontinuities, Boston: Kluwer Academic Publishers
Roy, J.R., Johansson, B. (1993): A Model of Trade Flows in Differentiated Goods. The Annals of Regional Science 27, 95-115
Ruelle, D. (1991): Chance and Chaos. Princeton: Princeton University Press
Ruffin, R.J. (1979): Growth and the Long-Run Theory of International Capital Movements. American Economic Review 69, 832-842
Salvadori, N. (1991): Post-Keynesian Theory of Distribution in the Long Run. In Nell, E.J., Semmler, W. (Eds.): Nicholas Kaldor and Mainstream Economics - Confrontation or Convergence?. London: Macmillan
Samuelson, P.A. (1965): A Theory of Induced Innovation Along Kennedy-Weizsäcker Lines. Review of Economics and Statistics XLVII, 343-56
Samuelson, P.A. (1971): Ohlin Was Right. Swedish Journal of Economics 73, 365-384
Sakashita, N. (1987) Optimum Location of Public Facilities Under the Influence of the Land Market. Journal of Regional Science 27, 1-12
Sasaki, K. (1990): The Establishment of a Subcenter and Urban Structure. Environment and Planning A 22, 369-383
Sato, K. (1966): The Neoclassical Theorem and Distribution of Income and Wealth. The Review of Economic Studies XXXIII, 331-336
Sato, R., Tsutsui, S. (1984): Technical Progress, the Schumpeterian Hypothesis and Market Structure. Journal of Economics S4, 1-37
Schmitz, J.A. (1989): Imitation, Entrepreneurship, and Long-Run Growth. Journal of Political Economy 97, 721-739
Schnare, A.B. (1976): Racial and Ethnic Price Differentials in an Urban Housing Market. Urban Studies 13, 107-20
Schultz, T.W. (1981): Investing in People - The Economics of Population Quality. Berkeley: University of California Press

Scotchmer, S., Thisse, J.F. (1992): Space and Competition - A Puzzle. Annals of Regional Science 26, 269-286

Senior, M.L., Wilson, A.G. (1974): Explorations and Sysnthesis of Linear Programming and Spatial Interaction Models of Residential Location. Geographical Analysis 6, 209-238

Shi, S.Y., Epstein, L.G. (1993): Habits and Time Preference. International Economic Review 34, 61-84

Simon, J.L., Love, D.O. (1990): City Size, Prices, and Efficiency for Individual Goods and Services. The Annals of Regional Science 24, 163-175

Shieh, Y.N. (1989): Demand, Location, and the Theory of Production. The Annals of Regional Science 23: 93-103

Shieh, Y.N., Mai, C.C. (1997): Demand and Location Decision of a Monopsonistic Firm. The Annals of Regional Science 31: 273-284

Sivitanidou, R., Wheaton, W.C. (1992): Wage and Rent Capitalization in the Commercial Real Estate Market. Journal of Urban Economics 31, 206-229

Smith, D. (1975): Neoclassical Growth Models and Regional Growth in the US'. Regional Science and Urban Economics 15, 165-181

Smith, J. (1977): Family Labor Supply over the Life Cycle. Explorations in Economic Research 4, 205-276

Smithies, A.F. (1941): Optimum Location in Spatial Competition. Journal of Political Economy 49: 423-39

Solow, R. (1956): A Contribution to the Theory of Growth. Quarterly Journal of Economics 70, 65-94

Solow, R.M. (1972): Congestion, Density, and the Use of Land in Transportation. The Swedish Journal of Economics 74, 161-173

Solow, R.M., Vickrey, W.S. (1971): Land Use in a Long Narrow City. Journal of Economic Theory 3, 430-447

Stahl, K. (1985): Existence of Equilibria in Spatial Economies. Regional Science and Urban Economics 15, 143-147

Stokey, N.L. (1991): The Volume and Composition of Trade Between Rich and Poor Countries. Review of Economic Studies 58, 63-80

Straszheim, M. (1987): The Theory of Urban Residential Location, in Handbook of Regional and Urban Economics. Volume II edited by E.S. Mills. Amsterdam: North-Holland

Stull, W.J. (1974) Land Use and Zoning in an Urban Economy. American Economic Review 64, 337-47

Suh, S.H. (1988): The Possibility and Impossibility of Intercity Commuting. Journal of Urban Economics 23, 86-100

Sweeney, J.L. (1974): Quality, Commodity Hierarchies and Housing Markets. Econometrica 42, 147-167

Swift, J.W. (1988): Hopf Bifurcation with the Symmetry of the Square. Nonlinearity 1, 333-377

Swan, T.W. (1956): Economic Growth and Capital Accumulation. Economic Record XXXII, 334-61

Takayama, T., Judge, G.G. (1971): Spatial and Temporal Price and Allocation Models. Amsterdam: North-Holland Publishing Company

Turing, A.M. (1952): The Chemical Basis of Morphogenesis. Phil. Trans. Roy. Soc. Lond. B. 237, 37-42

Turnovsky, S.J. (1997): International Macroeocnomic Dynamics. Mass., Cambridge: The MIT Press

Uzawa, H. (1965): Optimal Technical Change in an Aggregative Model of Economic Growth. International Economic Review 6: 18-31

Uzawa, H. (1968): Time Preference, the Consumption Function, and Optimum Asset Holdings, in Capital and Growth: Papers in Honour of Sir John Hicks edited by J.N. Wolfe. Chicago: Aldine

Voith, R. (1991): Capitalization of Local and Regional Attributes into Wages and Rents - Differences Across Residential, Commercial and Mixed-Use Communities. Journal of Regional Science 31, 129-145

von Thünen, J.H. (1826): Der Isolierte Staat in Beziehung auf Landwirtschaft und Nationalekonomie. Hamburg

Wan, H.Y. (1970): Optimal Savings Programs under Intertemporally Dependent Preferences. International Economic Review 11, 521-547

Wang, J.Y. (1990): Growth, Technology Transfer, and the Long-Run Theory of International Capital Movements. Journal of International Economics 29, 255-271

Wang, J.Y., Blomström, M. (1992): Foreign Investment and Technology Transfer - A Simple Model. European Economic Review 36, 137-155

Wang, P. (1993): Agglomeration in a Linear City with Heterogeneous Households. Regional Science and Urban Economics 23, 291-306

Weber, A. (1909): Alfred Weber's Theory of Location of Industries, translated from the German origin by C, J. Friedrich, 1929. Chicago: University of Chicago Press

Weidlich, W., Haag, G. (1983): Quantitative Sociology. Berlin: Springer

Weiss, Y., Willis, R.J. (1985): Children as Collective Goods and Divorce Settlements. Journal of Labor Economics 3, 268-292

Weizsäcker, C.C. (1966): Tentative Notes on a Two-Sector Model with Induced Technical Progress. Review of Economic Studies XXXIII, 245-51

Wheaton, W.C. (1979): Monocentric Models of Urban Land Use: Contributions and Criticism. In Mieszkowski, P., Straszheim, M. (Eds.): Current Issues in Urban Economics. Baltimore: Johns Hopkins University Press

Wigno, L., Jr. (1961): Transportation and Urban Land. Washington, DC.: Resources for the Future

Wilson, A.G. (1981): Catastrophe Theory and Bifurcation: Application to Urban and Regional Systems. London: Choom Helm.

Wilson, A. G. (1990): The Dynamics of Central Place Networks. In Chatterji, M., Kuenne, R.E. (Eds.) Dynamics and Conflict in Regional Structural Change. London: The Macmillan Press LTD

Yellin, J. (1974): Urban Population Distribution, Family Income, and Social Prejudice. Journal of Urban Economics 1, 21-47

Yinger, J. (1976): Racial Prejudice and Racial Residential Segregation in an Urban Model. Journal of Urban Economics 3, 383-96

Zhang, W.B. (1988a): The Pattern Formation of an Urban System. Geographical Analysis 20, 75-84

Zhang, W.B. (1988b): Urbanizing Processes With Moving Boundaries. Geographical Analysis 20, 328-339
Zhang, W.B. (1989a): Coexistence and Separation of Two Residential Groups - An Interactional Spatial Dynamic Approach. Geographical Analysis 21, 91-102
Zhang, W.B. (1989b): Urban Structural Changes in Continuous Time and Space. Umeå University Studies No. 191, University of Umeå
Zhang, W.B. (1989c): Spatial and Temporal Urban Pattern: Interpretations of the FitzHugh-Nagumo Equations. Umeå Economic Studies No. 190, University of Umeå
Zhang, W.B. (1990a): Economic Dynamics - Growth and Development. Heidelberg: Springer-Verlag
Zhang, W.B. (1990b): Stability versus Instability in Urban Pattern Formation. Socio-Spatial Dynamics 1, April, 41-56
Zhang, W.B. (1991a): Regional Dynamics with Creativity and Knowledge Diffusion. The Annals of Regional Science 25, 179-191
Zhang, W.B. (1991b): Economic Development with Creativity and Knowledge Diffusion. Socio-Spatial Dynamics 2, 19-30
Zhang, W.B. (1991c): Technological Change and Urban Growth - A Neoclassical Urban Economic Approach. Ricerche Economiche XLV, 591-608
Zhang, W.B. (1991d): Synergetic Economics. Berlin: Springer
Zhang, W.B. (1992a): A Two-Country Growth Model - Knowledge Accumulation with International Interactions. Journal of Scientific & Industrial Research 31, 187-194
Zhang, W.B. (1992b): Trade and World Economic Growth - Differences in Knowledge Utilization and Creativity. Economic Letters 39, 199-206
Zhang, W.B. (1993a): Location Choice and Land Use in an Isolated State - Endogenous Capital and Knowledge Accumulation. The Annals of Regional Science 27, 23-39
Zhang, W.B. (1993b): An Urban Pattern Dynamics with Capital and Knowledge Accumulation. Environment and Planning A, 25, 357-370
Zhang, W.B. (1993c): Knowledge, Economic Geography and Growth in a Two-Group Island-Economy, Technological Forecasting and Social Change 44, 405-418
Zhang, W.B. (1993d): Wages, Service Prices and Rent - Urban Division of Labor and Amenities. Seoul Journal of Economics 6, 97-113
Zhang, W.B. (1993e): Rent and Residential Distribution in a Two Group Neoclassical Growth Model. Economic Systems Research 5, 395-408
Zhang, W.B. (1994a): Capital, Population and Urban Patterns. Regional Science and Urban Economics 24, 273-286
Zhang, W.B. (1994b): Dynamics of Interacting Spatial Economies. Chaos, Solitons and Fractals 4, 595-604
Zhang, W.B. (1994c): Knowledge, Growth and Patterns of Trade. The Annals of Regional Science 28, 285-303
Zhang, W.B. (1994d): Dynamics of Economic Geography with Return to Scale. Working Paper of CERUM, University of Umeå, CWP-1994:2

Zhang, W.B. (1994e): Housing and Residential Location in a Two-Group Neoclassical Growth Model. Working Paper of CERUM, University of Umeå, CWP-1994:1

Zhang, W.B. (1995a): Leisure Time, Savings and Trade Patterns - A Two-Country Growth Model. Economic Modelling 12, 425-434

Zhang, W.B. (1995b): A Two-Country Dynamic Trade Model With Multiple Groups. International Economic Journal 9, 67-80

Zhang, W.B. (1996a): Knowledge and Value: Economic Structures wth Time and Space. Umeå: Umeå University

Zhang, W.B. (1996b): Growth, Trade and Land Values. Working Paper 96-1, Dept. of Social Systems Eng., Tottori University, Japan

Zhang, W.B. (1996c): Economic Growth and Urbanization - Amenity and Preference. Working Paper 96-2, Dept. of Social Systems Eng., Tottori University, Japan

Zhang, W.B. (1996a): Economic Growth, Housing and Residential Location. Umeå: Economic Studies No. 403, Umeå University

Zhang, W.B. (1997a): Economic Geography with Division of Labor and Amenity Difference. In Chatterji, M. (Ed.): Regional Science - Perspectives for the Future. London: Macmillan Press LTD

Zhang, W.B. (1997b): A Two-Region Model With Endogenous Capital and Knowledge - Locational Amenities and Preferences. International Review of Economics and Finance 6, 1-16

Zhang, W.B. (1998a): Economic Geography with Two Regions - Capital Accumulation and Economic Structure. Australian Economic Papers 35, 225-35

Zhang, W.B. (1998b): A Two-Region Growth Model - Competition, References, Resources, and Amenities. Papers in Regional Science 77, 173-188

Zhang, W.B. (1998c): Economic Growth with Pattern Formation and Preference Change. Geographical and Environmental Modelling 12, No.2

Zhang, W.B. (1999): Capital and Knowledge - Dynamics of Economic Structures with Non-Constant Returns. Heidelberg: Springer

Zhang, W.B. (2000): A Theory of International Trade – Capital, Knowledge and Economic Structures. Berlin: Springer

Author Index

Abdel-Rahman, H.M. 13
Aghion, P. 42
Allen, P. 186
Alonso, W. 5, 10, 25, 43, 61, 135, 151
Alperovich, G. 14, 15
Anas, A. 17, 24, 72, 120, 122
Anderson, P.W. 17
Andersson, Å.E. 13, 186, 202
Anderstig, C. 202
Arnott, R. 9, 24, 40, 72, 120, 122, 187, 202
Arrow, K.J. 17, 42, 105
Arthur, W.B. 186, 198
Asami, Y. 10
Ashwin, P. 195
Auer, L.von. 121

Bardhan, P.K. 167
Barro, R.J. 42, 202
Batten, D.F. 202
Becker, G.S. 61, 64, 79, 189
Beckmann, M.J. 4, 5, 9, 12, 61, 114, 118, 135, 151, 202
Bell, C. 114
Benson, B., 12
Berger, M.C. 150
Berliant, M. 10
Blaug, M. 1
Blomquist, G.C. 150
Blomström, M. 13
Boyer, M. 121
Brueckner, J.K. 24, 40, 72, 120, 187
Buiter, W.H. 166
Burmeister, E. 61, 76, 77, 94, 202

Calem, P.S. 136

Carlino, G.A. 136
Cheshire, P.C. 12
Chiappori, P.A. 64
Christaller, W. 1, 4
Cole, H.L. 121

David, P.A. 136
Dendrinos, D.S. 186, 187, 198
Devereux, M.B. 166
D'Aspremont, C. 11
Diamond, D.B. 10, 151
Dierx, A.H. 16
Dixit, A.K. 10, 15, 16, 202
Dixon, H.D.
Dobell, A.R. 61, 76, 77, 94, 202
Dollar, D. 42
Dosi, G. 42
Drandakis, E. 42

Eaton, B.C. 11, 166
Epstein, L.G. 121
Ethier, W.J. 13
Evans, A.W. 12

Fershtman, C. 121
Findlay, R. 166
Fisher, I. 125
Frenkel, J. 166
Friedman, B.M. 202
Fujita, M. 6, 9, 15, 21, 80, 106 135, 187
Fung, K.M. 13

Gabszewicz, J.J. 11
Gersovitz, M. 121, 125
Golubitsky, M. 195

Name Index

Greenhut, J. 12
Greenhut, M.L. 11,12, 135, 161, 202
Grossman, G.M. 42, 43, 94

Haag, G. 198
Haavelmo, T. 189, 202
Hahn, F.H. 203
Haken, H. 186
Hamilton, B.W. 80, 187
Hårsman, B. 202
Hartwick, J.M. 10, 15
Hartwick, P.G. 10, 15
Heckman, J.J. 64
Helpman, E. 42, 43, 94
Helsley, R.W. 15
Henderson, J.V. 12, 15, 21, 61, 76, 94, 114
Herbert, D.J. 77
Hockman, O. 24, 120, 187
Hoehn, J.C. 150
Hoover, E.M. Jr. 11
Hotelling, H. 11
Howitt, P. 42
Hung, C.S. 135, 151, 202

Ikeda, S. 166
Iooss, G. 193
Isard, W. 3,5, 135
Ishikawa, J. 13

Johansson, B. 42, 122, 202
Jonung, C. 61
Jones, R.W. 166
Joseph, D.D. 193
Judge, G.G. 202

Kanemoto, Y. 10, 26
Karlquist, A. 42, 122
Karlsson, C. 42
Kennedy, C. 42
Kern, C.R. 76
Kobayashi, K. 118
Krugman, P.R. 13, 21, 42

Lancaster, K. 202
Lipsey, R.G. 11

Lorenz, H.W. 17, 196
Lösch, A. 1,4, 15
Love, D.O. 150
Lucas, R.E. 14
Lundqvist, L. 122

Macurdy, T.E. 64
Mai, C.C. 4
Mailath, G.J. 121
Malthus, T.R. 202
Marshall, A. 14, 202
Mills, E.S. 10, 11, 15, 80, 187
Miyao, T. 40, 61, 76
Modigliani, F. 121, 125
Mohring, H. 5
Moses, L.M. 3
Mun, S.I. 202
Murry, J.D. 195
Muth, R.F. 5, 22, 24, 72, 77, 135

Nelson, R.R. 42
Newhouse, R.
Nicolis, G. 186
Niehans, J. 189, 202
Nijkamp, P. 17, 186
Norman, G. 3, 12, 135, 151, 202

Ogawa, H. 9, 15
Ohta, H. 12
Oniki, H. 166, 167
Ono, Y. 166

Palma, A.D. 61
Panico, C. 76
Papageorgiou, Y.Y. 5, 10, 61
Pasinetti, L.L. 76, 78
Pavitt, K. 42
Persson, I. 61
Phelps, E.S. 42
Phlips, L. 12
Pitchford, J.D. 189
Pines, D. 5,10, 24, 120, 187
Ponsard, C. 1
Postlewaite, A. 121
Prigogine, I. 186
Puu, T. 17, 151, 186, 202

Rabenau, B.V. 21, 40, 120, 187
Ram, R. 121, 125
Ramsey, F.
Rankin, N.
Rauch, J.E. 14, 114, 151, 166
Razin, A. 166, 167
Reggiani, A. 186
Rescher, N.
Richardson, H.W. 21, 76
Rivera-Batiz, F. 13
Roback, J. 114, 150
Robson, A.J. 42
Rodriguez, C.A. 166
Romer, P.M. 13, 42
Rose-Ackerman, S. 10, 76
Rosenbloom, J.L. 136
Rosser, J.B.Jr. 17, 186, 198
Roy, J.R. 122
Ruelle, D. 197
Ruffin, R.J. 166, 167

Sala-i-Martin, X. 42, 202
Salvadori, N. 76
Samuelson, P.A. 42, 166
Sakashita, N. 4
Sasaki, K. 151
Sanglier, M. 186
Sato, K. 76, 78, 105
Sato, R. 42
Schmitz, J.A. 13
Schnare, A.B. 76, 100
Schultz, T.W. 42
Scotchmer, S. 61, 150
Senior, M.L. 77
Shi, S.Y. 121, 166
Simon, J.L. 150
Shieh, Y.N. 4
Sivitanidou, R. 114
Smith, D. 21
Smith, J. 61
Smith, T.E. 10
Smithies, A.F. 11
Snickars, F. 122
Soete, L. 42
Solow, R.M. 10, 22, 62, 118

Sonis, M. 186, 198
Stahl, K. 10
Stengers, I. 186
Srevens, B.H. 77
Steward, I.N. 195
Stiglitz, J.E. 202
Stokey, N.L. 42
Stough, R. 17
Straszheim, M. 151, 187
Stull, W.J. 136
Suh, S.H. 118
Sullivan, A.M. 15
Sweeney, J.L. 187
Swift, J.W. 195
Swan, T.W. 22

Takayama, T. 202
Takens, F. 197
Ten Raa, T. 10
Thisse, J.F. 11, 61, 150
Tolley, G.S. 10, 151
Tsutsui, S. 42
Turing, A.M. 195
Turnovsky, S.J. 166

Uzawa, H. 42, 121, 166, 167

Verhoff, E. 17
Venables, A.J. 21
Vickrey, W.S. 9, 62, 118
Voith, R. 150
von Thünen, J.H. 1

Wan, H.Y. 121
Wang, J.Y. 13, 166
Wang, P. 40, 62
Weber, A. 1, 3, 12
Weidlich, W. 186, 198
Weiss, Y. 61, 121
Weizsäcker, C.C. 42
Wellisz, S.
Wheaton, W.C. 15, 114
Winter, S.G. 42
Wigno, L., Jr. 5, 135
Willis, R.J. 61
Wilson, A.G. 76, 77, 186

Yellin, J. 76, 100, 114
Yinger, J. 76, 100
Yoshikawa, K. 118

Zhang, W.B. 13, 17, 21, 25, 42, 43, 76, 77, 79, 94, 100, 114, 118, 120, 135, 151, 166, 167, 193, 196, 198, 202

Lecture Notes in Economics and Mathematical Systems

For information about Vols. 1–320
please contact your bookseller or Springer-Verlag

Vol. 321: M. Di Matteo, R.M. Goodwin, A. Vercelli (Eds.), Technological and Social Factors in Long Term Fluctuations. Proceedings. IX, 442 pages. 1989.

Vol. 322: T. Kollintzas (Ed.), The Rational Expectations Equilibrium Inventory Model. XI, 269 pages. 1989.

Vol. 323: M.B.M. de Koster, Capacity Oriented Analysis and Design of Production Systems. XII, 245 pages. 1989.

Vol. 324: I.M. Bomze, B.M. Pötscher, Game Theoretical Foundations of Evolutionary Stability. VI, 145 pages. 1989.

Vol. 325: P. Ferri, E. Greenberg, The Labor Market and Business Cycle Theories. X, 183 pages. 1989.

Vol. 326: Ch. Sauer, Alternative Theories of Output, Unemployment, and Inflation in Germany: 1960–1985. XIII, 206 pages. 1989.

Vol. 327: M. Tawada, Production Structure and International Trade. V, 132 pages. 1989.

Vol. 328: W. Güth, B. Kalkofen, Unique Solutions for Strategic Games. VII, 200 pages. 1989.

Vol. 329: G. Tillmann, Equity, Incentives, and Taxation. VI, 132 pages. 1989.

Vol. 330: P.M. Kort, Optimal Dynamic Investment Policies of a Value Maximizing Firm. VII, 185 pages. 1989.

Vol. 331: A. Lewandowski, A.P. Wierzbicki (Eds.), Aspiration Based Decision Support Systems. X, 400 pages. 1989.

Vol. 332: T.R. Gulledge, Jr., L.A. Litteral (Eds.), Cost Analysis Applications of Economics and Operations Research. Proceedings. VII, 422 pages. 1989.

Vol. 333: N. Dellaert, Production to Order. VII, 158 pages. 1989.

Vol. 334: H.-W. Lorenz, Nonlinear Dynamical Economics and Chaotic Motion. XI, 248 pages. 1989.

Vol. 335: A.G. Lockett, G. Islei (Eds.), Improving Decision Making in Organisations. Proceedings. IX, 606 pages. 1989.

Vol. 336: T. Puu, Nonlinear Economic Dynamics. VII, 119 pages. 1989.

Vol. 337: A. Lewandowski, I. Stanchev (Eds.), Methodology and Software for Interactive Decision Support. VIII, 309 pages. 1989.

Vol. 338: J.K. Ho, R.P. Sundarraj, DECOMP: An Implementation of Dantzig-Wolfe Decomposition for Linear Programming. VI, 206 pages.

Vol. 339: J. Terceiro Lomba, Estimation of Dynamic Econometric Models with Errors in Variables. VIII, 116 pages. 1990.

Vol. 340: T. Vasko, R. Ayres, L. Fontvieille (Eds.), Life Cycles and Long Waves. XIV, 293 pages. 1990.

Vol. 341: G.R. Uhlich, Descriptive Theories of Bargaining. IX, 165 pages. 1990.

Vol. 342: K. Okuguchi, F. Szidarovszky, The Theory of Oligopoly with Multi-Product Firms. V, 167 pages. 1990.

Vol. 343: C. Chiarella, The Elements of a Nonlinear Theory of Economic Dynamics. IX, 149 pages. 1990.

Vol. 344: K. Neumann, Stochastic Project Networks. XI, 237 pages. 1990.

Vol. 345: A. Cambini, E. Castagnoli, L. Martein, P Mazzoleni, S. Schaible (Eds.), Generalized Convexity and Fractional Programming with Economic Applications. Proceedings, 1988. VII, 361 pages. 1990.

Vol. 346: R. von Randow (Ed.), Integer Programming and Related Areas. A Classified Bibliography 1984–1987. XIII, 514 pages. 1990.

Vol. 347: D. Ríos Insua, Sensitivity Analysis in Multiobjective Decision Making. XI, 193 pages. 1990.

Vol. 348: H. Störmer, Binary Functions and their Applications. VIII, 151 pages. 1990.

Vol. 349: G.A. Pfann, Dynamic Modelling of Stochastic Demand for Manufacturing Employment. VI, 158 pages. 1990.

Vol. 350: W.-B. Zhang, Economic Dynamics. X, 232 pages. 1990.

Vol. 351: A. Lewandowski, V. Volkovich (Eds.), Multiobjective Problems of Mathematical Programming. Proceedings, 1988. VII, 315 pages. 1991.

Vol. 352: O. van Hilten, Optimal Firm Behaviour in the Context of Technological Progress and a Business Cycle. XII, 229 pages. 1991.

Vol. 353: G. Ricci (Ed.), Decision Processes in Economics. Proceedings, 1989. III, 209 pages 1991.

Vol. 354: M. Ivaldi, A Structural Analysis of Expectation Formation. XII, 230 pages. 1991.

Vol. 355: M. Salomon. Deterministic Lotsizing Models for Production Planning. VII, 158 pages. 1991.

Vol. 356: P. Korhonen, A. Lewandowski, J . Wallenius (Eds.), Multiple Criteria Decision Support. Proceedings, 1989. XII, 393 pages. 1991.

Vol. 357: P. Zörnig, Degeneracy Graphs and Simplex Cycling. XV, 194 pages. 1991.

Vol. 358: P. Knottnerus, Linear Models with Correlated Disturbances. VIII, 196 pages. 1991.

Vol. 359: E. de Jong, Exchange Rate Determination and Optimal Economic Policy Under Various Exchange Rate Regimes. VII, 270 pages. 1991.

Vol. 360: P. Stalder, Regime Translations, Spillovers and Buffer Stocks. VI, 193 pages . 1991.

Vol. 361: C. F. Daganzo, Logistics Systems Analysis. X, 321 pages. 1991.

Vol. 362: F. Gehrels, Essays in Macroeconomics of an Open Economy. VII, 183 pages. 1991.

Vol. 363: C. Puppe, Distorted Probabilities and Choice under Risk. VIII, 100 pages . 1991

Vol. 364: B. Horvath, Are Policy Variables Exogenous? XII, 162 pages. 1991.

Vol. 365: G. A. Heuer, U. Leopold-Wildburger. Balanced Silverman Games on General Discrete Sets. V, 140 pages. 1991.

Vol. 366: J. Gruber (Ed.), Econometric Decision Models. Proceedings, 1989. VIII, 636 pages. 1991.

Vol. 367: M. Grauer, D. B. Pressmar (Eds.), Parallel Computing and Mathematical Optimization. Proceedings. V, 208 pages. 1991.

Vol. 368: M. Fedrizzi, J. Kacprzyk, M. Roubens (Eds.), Interactive Fuzzy Optimization. VII, 216 pages. 1991.

Vol. 369: R. Koblo, The Visible Hand. VIII, 131 pages.1991.

Vol. 370: M. J. Beckmann, M. N. Gopalan, R. Subramanian (Eds.), Stochastic Processes and their Applications. Proceedings, 1990. XLI, 292 pages. 1991.

Vol. 371: A. Schmutzler, Flexibility and Adjustment to Information in Sequential Decision Problems. VIII, 198 pages. 1991.

Vol. 372: J. Esteban, The Social Viability of Money. X, 202 pages. 1991.

Vol. 373: A. Billot, Economic Theory of Fuzzy Equilibria. XIII, 164 pages. 1992.

Vol. 374: G. Pflug, U. Dieter (Eds.), Simulation and Optimization. Proceedings, 1990. X, 162 pages. 1992.

Vol. 375: S.-J. Chen, Ch.-L. Hwang, Fuzzy Multiple Attribute Decision Making. XII, 536 pages. 1992.

Vol. 376: K.-H. Jöckel, G. Rothe, W. Sendler (Eds.), Bootstrapping and Related Techniques. Proceedings, 1990. VIII, 247 pages. 1992.

Vol. 377: A. Villar, Operator Theorems with Applications to Distributive Problems and Equilibrium Models. XVI, 160 pages. 1992.

Vol. 378: W. Krabs, J. Zowe (Eds.), Modern Methods of Optimization. Proceedings, 1990. VIII, 348 pages. 1992.

Vol. 379: K. Marti (Ed.), Stochastic Optimization. Proceedings, 1990. VII, 182 pages. 1992.

Vol. 380: J. Odelstad, Invariance and Structural Dependence. XII, 245 pages. 1992.

Vol. 381: C. Giannini, Topics in Structural VAR Econometrics. XI, 131 pages. 1992.

Vol. 382: W. Oettli, D. Pallaschke (Eds.), Advances in Optimization. Proceedings, 1991. X, 527 pages. 1992.

Vol. 383: J. Vartiainen, Capital Accumulation in a Corporatist Economy. VII, 177 pages. 1992.

Vol. 384: A. Martina, Lectures on the Economic Theory of Taxation. XII, 313 pages. 1992.

Vol. 385: J. Gardeazabal, M. Regúlez, The Monetary Model of Exchange Rates and Cointegration. X, 194 pages. 1992.

Vol. 386: M. Desrochers, J.-M. Rousseau (Eds.), Computer-Aided Transit Scheduling. Proceedings, 1990. XIII, 432 pages. 1992.

Vol. 387: W. Gaertner, M. Klemisch-Ahlert, Social Choice and Bargaining Perspectives on Distributive Justice. VIII, 131 pages. 1992.

Vol. 388: D. Bartmann, M. J. Beckmann, Inventory Control. XV, 252 pages. 1992.

Vol. 389: B. Dutta, D. Mookherjee, T. Parthasarathy, T. Raghavan, D. Ray, S. Tijs (Eds.), Game Theory and Economic Applications. Proceedings, 1990. IX, 454 pages. 1992.

Vol. 390: G. Sorger, Minimum Impatience Theorem for Recursive Economic Models. X, 162 pages. 1992.

Vol. 391: C. Keser, Experimental Duopoly Markets with Demand Inertia. X, 150 pages. 1992.

Vol. 392: K. Frauendorfer, Stochastic Two-Stage Programming. VIII, 228 pages. 1992.

Vol. 393: B. Lucke, Price Stabilization on World Agricultural Markets. XI, 274 pages. 1992.

Vol. 394: Y.-J. Lai, C.-L. Hwang, Fuzzy Mathematical Programming. XIII, 301 pages. 1992.

Vol. 395: G. Haag, U. Mueller, K. G. Troitzsch (Eds.), Economic Evolution and Demographic Change. XVI, 409 pages. 1992.

Vol. 396: R. V. V. Vidal (Ed.), Applied Simulated Annealing. VIII, 358 pages. 1992.

Vol. 397: J. Wessels, A. P. Wierzbicki (Eds.), User-Oriented Methodology and Techniques of Decision Analysis and Support. Proceedings, 1991. XII, 295 pages. 1993.

Vol. 398: J.-P. Urbain, Exogeneity in Error Correction Models. XI, 189 pages. 1993.

Vol. 399: F. Gori, L. Geronazzo, M. Galeotti (Eds.), Nonlinear Dynamics in Economics and Social Sciences. Proceedings, 1991. VIII, 367 pages. 1993.

Vol. 400: H. Tanizaki, Nonlinear Filters. XII, 203 pages. 1993.

Vol. 401: K. Mosler, M. Scarsini, Stochastic Orders and Applications. V, 379 pages. 1993.

Vol. 402: A. van den Elzen, Adjustment Processes for Exchange Economies and Noncooperative Games. VII, 146 pages. 1993.

Vol. 403: G. Brennscheidt, Predictive Behavior. VI, 227 pages. 1993.

Vol. 404: Y.-J. Lai, Ch.-L. Hwang, Fuzzy Multiple Objective Decision Making. XIV, 475 pages. 1994.

Vol. 405: S. Komlósi, T. Rapcsák, S. Schaible (Eds.), Generalized Convexity. Proceedings, 1992. VIII, 404 pages. 1994.

Vol. 406: N. M. Hung, N. V. Quyen, Dynamic Timing Decisions Under Uncertainty. X, 194 pages. 1994.

Vol. 407: M. Ooms, Empirical Vector Autoregressive Modeling. XIII, 380 pages. 1994.

Vol. 408: K. Haase, Lotsizing and Scheduling for Production Planning. VIII, 118 pages. 1994.

Vol. 409: A. Sprecher, Resource-Constrained Project Scheduling. XII, 142 pages. 1994.

Vol. 410: R. Winkelmann, Count Data Models. XI, 213 pages. 1994.

Vol. 411: S. Dauzère-Péres, J.-B. Lasserre, An Integrated Approach in Production Planning and Scheduling. XVI, 137 pages. 1994.

Vol. 412: B. Kuon, Two-Person Bargaining Experiments with Incomplete Information. IX, 293 pages. 1994.

Vol. 413: R. Fiorito (Ed.), Inventory, Business Cycles and Monetary Transmission. VI, 287 pages. 1994.

Vol. 414: Y. Crama, A. Oerlemans, F. Spieksma, Production Planning in Automated Manufacturing. X, 210 pages. 1994.

Vol. 415: P. C. Nicola, Imperfect General Equilibrium. XI, 167 pages. 1994.

Vol. 416: H. S. J. Cesar, Control and Game Models of the Greenhouse Effect. XI, 225 pages. 1994.

Vol. 417: B. Ran, D. E. Boyce, Dynamic Urban Transportation Network Models. XV, 391 pages. 1994.

Vol. 418: P. Bogetoft, Non-Cooperative Planning Theory. XI, 309 pages. 1994.

Vol. 419: T. Maruyama, W. Takahashi (Eds.), Nonlinear and Convex Analysis in Economic Theory. VIII, 306 pages. 1995.

Vol. 420: M. Peeters, Time-To-Build. Interrelated Investment and Labour Demand Modelling. With Applications to Six OECD Countries. IX, 204 pages. 1995.

Vol. 421: C. Dang, Triangulations and Simplicial Methods. IX, 196 pages. 1995.

Vol. 422: D. S. Bridges, G. B. Mehta, Representations of Preference Orderings. X, 165 pages. 1995.

Vol. 423: K. Marti, P. Kall (Eds.), Stochastic Programming. Numerical Techniques and Engineering Applications. VIII, 351 pages. 1995.

Vol. 424: G. A. Heuer, U. Leopold-Wildburger, Silverman's Game. X, 283 pages. 1995.

Vol. 425: J. Kohlas, P.-A. Monney, A Mathematical Theory of Hints. XIII, 419 pages, 1995.

Vol. 426: B. Finkenstädt, Nonlinear Dynamics in Economics. IX, 156 pages. 1995.

Vol. 427: F. W. van Tongeren, Microsimulation Modelling of the Corporate Firm. XVII, 275 pages. 1995.

Vol. 428: A. A. Powell, Ch. W. Murphy, Inside a Modern Macroeconometric Model. XVIII, 424 pages. 1995.

Vol. 429: R. Durier, C. Michelot, Recent Developments in Optimization. VIII, 356 pages. 1995.

Vol. 430: J. R. Daduna, I. Branco, J. M. Pinto Paixão (Eds.), Computer-Aided Transit Scheduling. XIV, 374 pages. 1995.

Vol. 431: A. Aulin, Causal and Stochastic Elements in Business Cycles. XI, 116 pages. 1996.

Vol. 432: M. Tamiz (Ed.), Multi-Objective Programming and Goal Programming. VI, 359 pages. 1996.

Vol. 433: J. Menon, Exchange Rates and Prices. XIV, 313 pages. 1996.

Vol. 434: M. W. J. Blok, Dynamic Models of the Firm. VII, 193 pages. 1996.

Vol. 435: L. Chen, Interest Rate Dynamics, Derivatives Pricing, and Risk Management. XII, 149 pages. 1996.

Vol. 436: M. Klemisch-Ahlert, Bargaining in Economic and Ethical Environments. IX, 155 pages. 1996.

Vol. 437: C. Jordan, Batching and Scheduling. IX, 178 pages. 1996.

Vol. 438: A. Villar, General Equilibrium with Increasing Returns. XIII, 164 pages. 1996.

Vol. 439: M. Zenner, Learning to Become Rational. VII, 201 pages. 1996.

Vol. 440: W. Ryll, Litigation and Settlement in a Game with Incomplete Information. VIII, 174 pages. 1996.

Vol. 441: H. Dawid, Adaptive Learning by Genetic Algorithms. IX, 166 pages.1996.

Vol. 442: L. Corchón, Theories of Imperfectly Competitive Markets. XIII, 163 pages. 1996.

Vol. 443: G. Lang, On Overlapping Generations Models with Productive Capital. X, 98 pages. 1996.

Vol. 444: S. Jørgensen, G. Zaccour (Eds.), Dynamic Competitive Analysis in Marketing. X, 285 pages. 1996.

Vol. 445: A. H. Christer, S. Osaki, L. C. Thomas (Eds.), Stochastic Modelling in Innovative Manufactoring. X, 361 pages. 1997.

Vol. 446: G. Dhaene, Encompassing. X, 160 pages. 1997.

Vol. 447: A. Artale, Rings in Auctions. X, 172 pages. 1997.

Vol. 448: G. Fandel, T. Gal (Eds.), Multiple Criteria Decision Making. XII, 678 pages. 1997.

Vol. 449: F. Fang, M. Sanglier (Eds.), Complexity and Self-Organization in Social and Economic Systems. IX, 317 pages, 1997.

Vol. 450: P. M. Pardalos, D. W. Hearn, W. W. Hager, (Eds.), Network Optimization. VIII, 485 pages, 1997.

Vol. 451: M. Salge, Rational Bubbles. Theoretical Basis, Economic Relevance, and Empirical Evidence with a Special Emphasis on the German Stock Market.IX, 265 pages. 1997.

Vol. 452: P. Gritzmann, R. Horst, E. Sachs, R. Tichatschke (Eds.), Recent Advances in Optimization. VIII, 379 pages. 1997.

Vol. 453: A. S. Tangian, J. Gruber (Eds.), Constructing Scalar-Valued Objective Functions. VIII, 298 pages. 1997.

Vol. 454: H.-M. Krolzig, Markov-Switching Vector Autoregressions. XIV, 358 pages. 1997.

Vol. 455: R. Caballero, F. Ruiz, R. E. Steuer (Eds.), Advances in Multiple Objective and Goal Programming. VIII, 391 pages. 1997.

Vol. 456: R. Conte, R. Hegselmann, P. Terna (Eds.), Simulating Social Phenomena. VIII, 536 pages. 1997.

Vol. 457: C. Hsu, Volume and the Nonlinear Dynamics of Stock Returns. VIII, 133 pages. 1998.

Vol. 458: K. Marti, P. Kall (Eds.), Stochastic Programming Methods and Technical Applications. X, 437 pages. 1998.

Vol. 459: H. K. Ryu, D. J. Slottje, Measuring Trends in U.S. Income Inequality. XI, 195 pages. 1998.

Vol. 460: B. Fleischmann, J. A. E. E. van Nunen, M. G. Speranza, P. Stähly, Advances in Distribution Logistic. XI, 535 pages. 1998.

Vol. 461: U. Schmidt, Axiomatic Utility Theory under Risk. XV, 201 pages. 1998.

Vol. 462: L. von Auer, Dynamic Preferences, Choice Mechanisms, and Welfare. XII, 226 pages. 1998.

Vol. 463: G. Abraham-Frois (Ed.), Non-Linear Dynamics and Endogenous Cycles. VI, 204 pages. 1998.

Vol. 464: A. Aulin, The Impact of Science on Economic Growth and its Cycles. IX, 204 pages. 1998.

Vol. 465: T. J. Stewart, R. C. van den Honert (Eds.), Trends in Multicriteria Decision Making. X, 448 pages. 1998.

Vol. 466: A. Sadrieh, The Alternating Double Auction Market. VII, 350 pages. 1998.

Vol. 467: H. Hennig-Schmidt, Bargaining in a Video Experiment. Determinants of Boundedly Rational Behavior. XII, 221 pages. 1999.

Vol. 468: A. Ziegler, A Game Theory Analysis of Options. XIV, 145 pages. 1999.

Vol. 469: M. P. Vogel, Environmental Kuznets Curves. XIII, 197 pages. 1999.

Vol. 470: M. Ammann, Pricing Derivative Credit Risk. XII, 228 pages. 1999.

Vol. 471: N. H. M. Wilson (Ed.), Computer-Aided Transit Scheduling. XI, 444 pages. 1999.

Vol. 472: J.-R. Tyran, Money Illusion and Strategic Complementarity as Causes of Monetary Non-Neutrality. X, 228 pages. 1999.

Vol. 473: S. Helber, Performance Analysis of Flow Lines with Non-Linear Flow of Material. IX, 280 pages. 1999.

Vol. 474: U. Schwalbe, The Core of Economies with Asymmetric Information. IX, 141 pages. 1999.

Vol. 475: L. Kaas, Dynamic Macroeconomics with Imperfect Competition. XI, 155 pages. 1999.

Vol. 476: R. Demel, Fiscal Policy, Public Debt and the Term Structure of Interest Rates. X, 279 pages. 1999.

Vol. 477: M. Théra, R. Tichatschke (Eds.), Ill-posed Variational Problems and Regularization Techniques. VIII, 274 pages. 1999.

Vol. 478: S. Hartmann, Project Scheduling under Limited Resources. XII, 221 pages. 1999.

Vol. 479: L. v. Thadden, Money, Inflation, and Capital Formation. IX, 192 pages. 1999.

Vol. 480: M. Grazia Speranza, P. Stähly (Eds.), New Trends in Distribution Logistics. X, 336 pages. 1999.

Vol. 481: V. H. Nguyen, J. J. Strodiot, P. Tossings (Eds.). Optimation. IX, 498 pages. 2000.

Vol. 482: W. B. Zhang, A Theory of International Trade. XI, 192 pages. 2000.

Vol. 483: M. Königstein, Equity, Efficiency and Evolutionary Stability in Bargaining Games with Joint Production. XII, 197 pages. 2000.

Vol. 484: D. D. Gatti, M. Gallegati, A. Kirman, Interaction and Market Structure. VI, 298 pages. 2000.

Vol. 485: A. Garnaev, Search Games and Other Applications of Game Theory. VIII, 145 pages. 2000.

Vol. 486: M. Neugart, Nonlinear Labor Market Dynamics. X, 175 pages. 2000.

Vol. 487: Y. Y. Haimes, R. E. Steuer (Eds.), Research and Practice in Multiple Criteria Decision Making. XVII, 553 pages. 2000.

Vol. 488: B. Schmolck, Ommitted Variable Tests and Dynamic Specification. X, 144 pages. 2000.

Vol. 489: T. Steger, Transitional Dynamics and Economic Growth in Developing Countries. VIII, 151 pages. 2000.

Vol. 490: S. Minner, Strategic Safety Stocks in Supply Chains. XI, 214 pages. 2000.

Vol. 491: M. Ehrgott, Multicriteria Optimization. VIII, 242 pages. 2000.

Vol. 492: T. Phan Huy, Constraint Propagation in Flexible Manufacturing. IX, 258 pages. 2000.

Vol. 493: J. Zhu, Modular Pricing of Options. X, 170 pages. 2000.

Vol. 494: D. Franzen, Design of Master Agreements for OTC Derivatives. VIII, 175 pages. 2001.

Vol. 495: I Konnov, Combined Relaxation Methods for Variational Inequalities. XI, 181 pages. 2001.

Vol. 496: P. Weiß, Unemployment in Open Economies. XII, 226 pages. 2001.

Vol. 497: J. Inkmann, Conditional Moment Estimation of Nonlinear Equation Systems. VIII, 214 pages. 2001.

Vol. 498: M. Reutter, A Macroeconomic Model of West German Unemployment. X, 125 pages. 2001.

Vol. 499: A. Casajus, Focal Points in Framed Games. XI, 131 pages. 2001.

Vol. 500: F. Nardini, Technical Progress and Economic Growth. XVII, 191 pages. 2001.

Vol. 501: M. Fleischmann, Quantitative Models for Reverse Logistics. XI, 181 pages. 2001.

Vol. 502: N. Hadjisavvas, J. E. Martínez-Legaz, J.-P. Penot (Eds.), Generalized Convexity and Generalized Monotonicity. IX, 410 pages. 2001.

Vol. 503: A. Kirman, J.-B. Zimmermann (Eds.), Economics with Heterogenous Interacting Agents. VII, 343 pages. 2001.

Vol. 504: P.-Y. Moix (Ed.),The Measurement of Market Risk. XI, 272 pages. 2001.

Vol. 505: S. Voß, J. R. Daduna (Eds.), Computer-Aided Scheduling of Public Transport. XI, 466 pages. 2001.

Vol. 506: B. P. Kellerhals, Financial Pricing Models in Continuous Time and Kalman Filtering. XIV, 247 pages. 2001.

Vol. 507: M. Koksalan, S. Zionts, Multiple Criteria Decision Making in the New Millenium. XII, 481 pages. 2001.

Vol. 508: K. Neumann, C. Schwindt, J. Zimmermann, Project Scheduling with Time Windows and Scarce Resources. XI, 335 pages. 2002.

Vol. 509: D. Hornung, Investment, R&D, and Long-Run Growth. XVI, 194 pages. 2002.

Vol. 510: A. S. Tangian, Constructing and Applying Objective Functions. XII, 582 pages. 2002.

Vol. 511: M. Külpmann, Stock Market Overreaction and Fundamental Valuation. IX, 198 pages. 2002.

Vol. 512: W.-B. Zhang, An Economic Theory of Cities.XI, 220 pages. 2002.